Praise for *Artificial Intelligence: A Beginner's Guide*

"I often meet people who have recently discovered the thrill of artificial intelligence and want to know more, and occasionally I come across those who are confused, worried, or grossly misinformed about it. I've long wished there was a clear, concise, comprehensive and infectiously enthusiastic text that I could point to and say, 'read this and you will understand.' This is that book. Read it and you will understand."

Steve Grand, author of *Creation: Life and How to Make it*

"Blay Whitby's book is exciting, informative, up-to-date, and refreshingly sane. He manages to give both a rounded and a pointed picture of an area of science and technology that's affecting our lives already. This book is a 'must' for anyone who wants to find out about AI. It may make you gasp. For sure, it will make you think."

Margaret A. Boden, University of Sussex

"*Artificial Intelligence: A Beginner's Guide* is a clear and level-headed approach to the subject. Unlike some other beginner's guides and popular books on the subject it avoids the hype about how AI will take over the human race and robots will one day rule the planet. Instead, it provides a readable account of where AI research is today and makes it clear what AI is not."

Professor Noel Sharkey, University of Sheffield

# artificial

# intelligence

## a beginner's guide

blay whitby

ONEWORLD

OXFORD

# ARTIFICIAL INTELLIGENCE: A BEGINNER'S GUIDE

Oneworld Publications
(Sales and Editorial)
185 Banbury Road
Oxford OX2 7AR
England
www.oneworld-publications.com

ISBN 1–85168–322–4

Cover design by the Bridgewater Book Company
Typeset by Saxon Graphics Ltd, Derby, UK
Illustrations on pp. 45, 47 and 60 by Deirdre Counihan
Printed and bound in Spain by Book Print S.L.

*For Sharon Wood for introducing me to the field of Artificial Intelligence*

# contents

# acknowledgements

Many more people than one could ever thank contribute to the writing of a book such as this. I am, for example, indebted to countless students whose constant need for clear and concise explanations has, in that great paradox of teaching, taught me so much about my material. Similarly I cannot possibly name all the colleagues who have helped either through challenging or supporting my ideas over the years.

I am indebted to Ann Grand for pointing out that daisies are programmed to take over the world and to the following friends and colleagues who were kind enough to comment on portions of the text: Maggie Boden, Olivia Boyer, Ron Chrisley, Andy Clark, Rob Clowes, Deirdre Counihan, Kyran Dale, Steve Draper, Carlisle George, Dave Nicholson, Mike Sharples, Aaron Sloman, Nick Szczepanik, Steve Torrance, Des Watson, Emily Whitby, Sharon Wood, and David Young. Many helpful comments were also provided by the publisher's anonymous reader. I must confess that I did not always heed their advice and so responsibility for the various inadequacies of the book must remain with me.

# preface

This book is intended to introduce the reader to the fascinating world of Artificial Intelligence. Like the other titles in the Oneworld series, it presumes no previous knowledge of the area but assumes that both the subject and the excitement around it can be conveyed to a general audience in a non-technical and jargon-free manner.

Artificial Intelligence (normally abbreviated to AI) is one of the most – possibly *the most* – exciting challenges that humanity has ever undertaken, although it is hard perhaps to convey excitement within the pages of a relatively serious book. However, the unrestrained gusto with which researchers from the widest possible variety of backgrounds throw themselves into the many diverse challenges of AI makes it a very exciting place to work. Some of the challenges described may, at first glance, seem over-ambitious but reaching for the stars and ignoring the cynics is characteristic of a young science. I have been writing on AI for twenty years and have no illusions as to just how difficult some of its goals are to attain. However, I consider it an inestimable privilege to have been in the same room as people who, on being told that something was impossible, promptly went out and did it.

AI is not only all about ambitious remote goals: it is also a remarkably successful technology. Much of what was once presented as cutting-edge ideas is now regarded as everyday computing technology. Ex-students and colleagues who have entered the commercial sector have, on occasion, made small fortunes from developing and exploiting AI technologies.

There is no shortage of good books on AI. However, few of them are particularly suited to introduce a beginner to the field. Student texts have to deal with the subject in a more technical manner and fluency with various types of formal notation is usually required.

AI is an extremely diverse area and it would be impossible for this book (or possibly any book) to cover every important piece of research and technology in detail. In order to introduce the general reader to as much of the area as possible I have inevitably had to be selective and to describe some of the technical details rather briefly. The bibliography should allow the curious reader to fill in any missing details.

Because this is a book for the general reader I have refrained from introducing any equations, computer code, logic, or symbolic descriptions. I have also kept scholarly references to a minimal level. In this sense, it is very different from an AI textbook – of which there are many excellent examples available. My personal favourites are listed in the "further reading" sections for the relevant chapters (including this preface) but omission of a title does not imply any criticism.

This book alone, therefore, cannot prepare you for a job in AI. If you are interested in pursuing a career in AI you should, if at all possible, learn some computer programming. This is not because AI researchers spend all their time writing computer programs – in fact they tend to spend most of their time thinking about difficult problems – it is simply because computer programming is a fundamental skill in this area and the closest thing there is to a "lingua franca" in AI.

## two myths about ai

There are two particularly pernicious myths about AI which are in general circulation and which this book will comprehensively dismiss. The first myth is the claim that AI has failed. (Some people occasionally make the stronger claim that AI is impossible.) This myth is just plain false. AI has succeeded in producing a steady stream of successful technologies that are changing the way we live. Examples of the successes of AI litter the chapters of this book. There can be little doubt that the activities of AI researchers will continue to spin off useful everyday technologies. AI has also succeeded in producing a large number of important ideas and methods that have spectacularly influenced other areas of science and art.

Why such a myth might persist probably lies in a misperception of the goals of AI. It is easy to become confused about these goals since AI has and always has had multiple goals – ranging from making computers easier to use, to understanding human thought. There have always been those who expressed cynicism about AI – both inside and outside the field. Fluidity of goals and methods is typical of, and probably helpful to, an area of science that is still very young.

The second myth is the notion that AI, "when it succeeds", will leave humanity in a state of serfdom to a new race of all-powerful machines. This second myth is not merely false: it is also nonsensical. It must be admitted that some of the responsibility for the introduction of this myth lies with certain individuals within AI. Even so, it makes no sense. There is nothing in present-day or foreseeable AI that is likely to bring about this dystopian prospect. AI has, so far, proved to be a remarkably benign technology. There is, as with all technologies, a balance of costs and benefits which we will consider in detail. However, AI compares very favourably with many other technologies of the second half of the twentieth century in this respect.

## the structure of the book

Many, probably most, books on AI are partial in the sense of advocating particular approaches and ignoring or denigrating others. This would be inappropriate in a beginner's guide, which should give a broad brush picture of the entire area. It is inevitable that my own perspective permeates the text but I have tried to be catholic and eclectic throughout. For similar reasons, I have avoided presenting the subject in an historical fashion, which would inevitably give the largely false impression that the latest fashions had replaced earlier techniques. Actually, AI has not progressed in such a linear fashion. Techniques and ideas tend to be in the limelight for up to a decade, and are then replaced, only to be "rediscovered" and come back into fashion again a couple of decades later.

The first chapter deals with the difficult question of what AI is, outsiders naturally tending to assume that just one view of the subject prevails within it. This is not the case for there is widespread disagreement within AI – not least about the proper scope and goals of the subject. Agonizing over the definition of an area of study, particularly by someone who has taught and written about it for years, may seem unforgivably self-indulgent. However, it is important to remember that definitions are crucial to progress in science and an infant science like AI provides plenty of scope for many different views and techniques.

Chapter 2 looks at AI at work. Various successful applications of AI are examined and their workings explained. Chapter 3 introduces the many areas of AI research and applications which are inspired by biology in various ways. This includes attempts to design and build computer programs which are more like brains in their organization and the development of programs which imitate biological evolution. In chapter 4 we step back from the subject a bit to look at some of the big challenges that still remain for AI and at some current strands of research that are attempting to meet these challenges.

Under the heading "AI Diffuses", chapter 5 examines the wider impacts of AI. As a science AI has influenced many other areas of science and art through the export of powerful and productive ideas. Nowhere is this more obvious than in the development of cognitive science. Largely inspired by AI ideas and technologies, this new area of science has grown spectacularly. At my own university, the last twenty-five years have seen cognitive science grow from the part-time activity of half a dozen researchers from various backgrounds to become the largest science subject in the university.

Chapter 6 considers the social impacts of AI. As a technology, AI has already had some degree of social impact and we may expect more. We are entering an age in which we will rely on smart machines to assist us in many intellectual tasks. People may fear redundancy and decadence but there is every prospect that we will use these machines as a springboard to even greater human achievement than we can now envisage. It is indeed a privilege to be given the opportunity to write about such exciting trends.

Blay Whitby
University of Sussex
February 2003

## further reading

A good general AI textbook is *Artificial Intelligence, A Modern Approach* (Russell and Norvig, 2003). Since it adopts an agent-based approach to AI, it is an excellent starting point in pursuing many of the ideas in this book.

A less technical book about many early AI programs is *Artificial Intelligence and Natural Man* by Margaret Boden (1987).

Should you decide to learn computer programming, it's probably best to learn by doing rather than from a book.

# what is ai?

## what ai is

Artificial Intelligence (AI) is the study of intelligent behaviour (in humans, animals, and machines) and the attempt to find ways in which such behaviour could be engineered in any type of artifact. It is one of the most difficult and arguably the most exciting enterprise ever undertaken by humanity.

The difficulty of the enterprise may not be obvious at first, but it has been painfully learned. Sometimes the quest for AI is likened to deep space exploration in its level of difficulty. However, the truth is that it is much harder than that. In space exploration we have, at least, an understanding of the main technical problems. In AI, unfortunately, that is not yet the case.

On the other hand, the difficulty is more than compensated for by the potential rewards, both practical and intellectual. In practical terms, AI has already more than justified itself. As we shall see in chapter 2, applications of AI and spin-offs from AI are already shaping our technology and society and will increasingly do so in the future.

The intellectual rewards are even more exciting. AI offers (and is now beginning to provide) a scientific understanding of some of the most difficult questions we could ever pose about ourselves and our world. We are at the start of a journey into the most challenging "inner space" – fundamental questions about what it is to be a thinking being at all.

Many people find such a scientific quest unnerving. It is unnerving. The possibility that some of our most cherished human attributes might be scientifically explained could be seen as a sort of threat. It is important not to make this threat into some sort of bogeyman that it is not. As we shall see, there are many aspects of human thought that AI has not yet investigated and it may well be that they remain forever uninvestigated. Secondly, the fact that we have, for example, a scientific account of the formation of a rainbow does not detract from its beauty. If we were at some point in the future to provide a scientific account of the working of human creativity, that would not make any of the products of that creativity less beautiful or less interesting. The wonderful products of the human mind are undiminished by explanation.

The real root of the threat is probably a predilection for mystery. Humans tend to like a bit of mystery in their stories about the world and particularly like a bit of mystery in how they take their most cherished decisions and come up with their best ideas. At the same time, though, humans are driven to explore. Just as we must constantly strive to find out about the limits of the universe, so we must constantly probe the details of how our intelligence, and that of other animals, works. Intelligence and its many wonderful products are not threatened by any such enquiry. On the contrary, the surprising difficulties involved in getting some apparently simple aspects of intelligent behaviour from machines can only inspire awe at just how wonderful natural intelligence is.

Another important consequence of the definition of AI given at the start of this section is the way in which it clearly transcends conventional boundaries of study. It is both science and engineering, since it involves the study of and the building of intelligent behaviour. Indeed work in AI often tends to make a mockery of the supposed division between science and engineering, as it involves finding out through building. Even more radical is the fact that "intelligent behaviour" is to be found in many diverse places. It is in the communication between bees, the movement of prices on the stock exchange, the use of metaphor in Hamlet, and in the working of an automated air traffic control system. If we wish to understand it we must be prepared to pursue it through all these areas and more. This makes complete nonsense of traditional boundaries between arts and science, between engineering and biology, between the individual and society.

AI has always been a truly interdisciplinary enterprise, being at the same time both art and science, both engineering and psychology. If this claim seems too extravagant at the abstract level, you should be aware that AI has already produced programs that simulate the rantings of a paranoid schizophrenic, or the birth, breeding, and evolution of synthetic creatures. It has produced programs that have discovered new mathematical theorems and programs that can improvise jazz. It has produced programs that can detect fraudulent financial transactions and robots that can clear away empty coke tins from a laboratory. There are programs that paint pictures, programs that perform medical diagnosis, programs that teach, and programs that learn.

This is not an attempt to give the impression that AI has already conquered all areas of knowledge. Far from it; these successes are local, cranky, and fail to generalize across other areas. The best that can be said is that they are tantalizing little glimpses of what might one day be achieved. Most importantly, they are evidence of the almost incredible breadth of vision that inspires research in AI.

## what ai is not

The first, and perhaps the most important, step in understanding AI is to abandon one's preconceptions. Most readers will have already formed some hazy ideas of what AI is about. These hazy ideas will be almost completely wrong. As far as possible, you should dismiss them before continuing further.

You will, for example, have some sort of understanding of the term "intelligence". This might lead you to assume that artificial intelligence has something to do with building the sort of intelligence which you possess into some sort of machine. However, as we shall see repeatedly throughout this book, that assumption can be highly misleading. Work in artificial intelligence keeps on showing that we do not understand our own intelligence in any scientific way. Even more surprising is how it has revealed that the variety of ways in which humans use their intelligence to solve problems are certainly not the only ways available and often they are not the best.

There are some good reasons (discussed in chapters 3 and 4) for believing that the study of *human* intelligence is often not helpful in AI. Not only do we lack scientific understanding of most of the relevant details of human intelligence but it would be way beyond the state of the art to try to imitate it in machines. Many AI researchers look at supposedly simpler animals, such as insects, believing that human intelligence is just too complex to inform their work at all.

On the other hand, other AI researchers have achieved spectacular results in getting particular aspects of human behaviour from machines. A good example would be chess playing. Certainly, when work started in this area in the 1950s it was seen as a good example of intelligent human behaviour. In a tournament in 1997 a computer called Deep Blue beat the World Chess Champion, Gary Kasparov. That machines can now play chess very well has been established.

However, when we look in detail (in chapter 2) at the way in which computers play chess we shall see that it is probably very different from the ways in which humans play chess. It might seem a little provocative to say that computers play chess better than humans do but, in this field, better means winning. Beating the human World Champion is as much as we can reasonably ask of a chess-playing program. In at least this one area, it seems that we can justifiably talk of having found better methods than those used by humans.

Another set of preconceptions about AI comes from the world of science fiction. Intelligent machines, robots, cyborgs, and so on are favourite themes of all forms of science fiction. Unfortunately once again, the impressions gained from science fiction are highly misleading. It is important to remember that science fiction is essentially just that – fiction. It has very often inspired scientists in AI and other areas, but it may give a mistaken picture of what is actually happening in current research. In particular, science fiction may lead readers to assume that far more has been achieved than is actually the case and that AI is more human-like than is actually the case. This book will challenge these assumptions.

The final set of preconceptions comes from some pervasive myths about computers. These myths are so widely believed, even by some computer scientists, that it requires courage to challenge them, but challenge them we must. It is often said that "computers can do only what we tell them to do". Like all good myths, this one has an element of truth. All computers need elaborate software (usually programs written by humans) in order to operate at all. However if this slogan is read as "all that computers ever do is to follow explicit instructions" then it is very wrong. In this book readers will be introduced to programs that take guesses, play hunches, and in many ways out-perform their creators. We will hear about robots that have not been designed, but rather have been evolved.

For similar reasons, readers are implored to relinquish the myth that computers are purely rational, deductive machines. AI researchers have not confined their study and experimentation to the rational parts of intelligent behaviour. It is true that greater success has been achieved in the obviously more rational areas, but AI has also involved getting computers to do some very non-rational things. One discovery of research in AI is that whole areas of intelligent behaviour have nothing to do with logic and deduction. Some different methods are needed, but there have been surprising successes in getting computers to perform in these areas too.

## ai methods and tools

AI researchers have never defined their area in terms of one particular set of research methods. We might say that in AI there are at least as many methods as there are researchers. There are a number of reasons for this. First, such a wide-ranging and cross-disciplinary enterprise simply has to be eclectic in its choice of methods.

Second it has always been easy for individual AI researchers to pursue multiple goals. Consider, for example, the field of AI known as NLP (natural language processing). This involves the study of and attempt to build computer programs with which we might communicate in English or other human languages. It has the simultaneous goals of (at least) making computers easier for us to use, understanding the complex rules that govern the structure of natural languages, and of discovering just how humans can learn and apply those rules. Different researchers may emphasize each of these three goals differently. Indeed the same researcher may well give different goals prominence on different occasions, depending on the audience.

What is true of the area of NLP is true of AI as a whole. It is a study that has always had multiple goals. To extend a crude military metaphor often used in science, we could say that AI has chosen to attack its problem area on the widest possible front. Instead of a concentrated attack that might lead to the core of the problem of intelligence, there are innumerable little skirmishes, along a line that stretches across most of human knowledge. Some of these skirmishes are going well and some not so well, but over a period of a few years that can change. Perhaps understandably, those researchers whose skirmish seems to be yielding results tend to shout that this is the long awaited breakthrough. When the dust settles, however, it becomes clear that once again they have only moved the frontline forward a few yards. At the same time, other researchers may give very plausible reasons why their particular part of the line is where the big push should be made. In the history of AI, however, such "big pushes" have only resulted in moving the line forward locally a few hundred yards. So far, there is steady progress in AI, but no big breakthroughs.

The main tool used in AI research is the digital computer. This most certainly does not mean that AI is about digital computers. The computer is, firstly, a tool. It is used because it enables researchers quickly to build "models of behaviour" and examine them. Some AI researchers feel that it is necessary to build real robots that interact with the real world. We shall look in detail at the reasons why they believe this in a later chapter. For now, it should suffice to say that most AI research makes extensive use of computer programs that model aspects of the real world.

There are a number of difficult issues raised by this method of research. Many readers will intuitively feel that there is a world of difference between simulated thought and "real thought" and we will go into more detail on this later in the book. For now, it should be sufficient to say that in many other areas this use of computers is very effective. We would expect civil engineers, for example, to make use of computer simulation to discover how a bridge might withstand hurricane force winds. To build a real bridge and wait for "once in a century" weather conditions would be stupid. The digital computer can answer the civil engineers' questions in a matter of hours. It is the ability of the computer to work through myriad possibilities in a very short time that makes it such a useful tool in AI.

Modern computers allow us to build detailed working models of things far more complex than bridges and winds. Just as we can now store pictures and music in digital format, so we can work with digital manipulations of anything that we can describe accurately enough. This need for accurate description is, perhaps, the key to understanding AI research methods. The second important role of the computer (or physical robot) in AI is to prompt researchers to ask certain specific questions about natural intelligence. The computer is used not only to simulate, as with the bridge, but also to inspire certain scientific questions. Asking how we might get the computer to do it imposes a new scientific rigour on our way of looking at familiar things.

An analogy is often made between AI and "artificial flight". It is an analogy that can be usefully deployed here. At the beginning of the twentieth century there was a very limited understanding of the way in which birds and insects were able to fly: it was obvious that they did and the scientific enquiry tended to end at that point. Indeed, at the time when the Wright brothers started the aviation age (in 1903) most biology textbooks said that birds could fly "because they had the power of flight". This is an account that owes much to Aristotle, writing in Athens in the fourth century BC. However, it is not much help to those who wish to understand how to build aircraft. As a result of the more detailed understanding of flight which stemmed from the building of successful aircraft, we now know that birds fly by complying with the laws of aerodynamics. The quest for some sort of analogous "aerodynamics of intelligence" is often seen as the ultimate goal of AI.

Just as scientific understanding of bird flight stemmed from the building of aircraft so, it is hoped, scientific understanding of intelligence will stem from attempting to build intelligent machines.

Of course, AI research involves far more than just writing computer programs. If you wish to build something resembling the intelligence of a particular animal then it behoves you to study that animal in great detail. Much AI research involves doing a sort of biology or a sort of psychology, and even a sort of philosophy. Ignoring fatuous "turf wars" between disciplines (and such wars deserve nothing more), we can see that the use of the computer as a tool makes these "sorts" more rigorous than they would be without it.

An illustrative example is not hard to find. Take the task which you are now performing: reading. Obviously you have "the power of reading" to have got this far. That you have reasonably clear printing to follow and at least one suitable eye (or fingertip, if you read in Braille), we might also deduce. However, these truths would not be of much help in the design of a reading machine. If we wish to build a machine that can read we need to ask some more detailed questions.

Are you, for example, looking at each individual letter and comparing it to a library of characters (at least 114 in number) to determine what it is, before proceeding to the next? On reaching a blank space and assembling the letters into a word, do you then search through some sort of dictionary (a much larger number is involved here) in order to recognize it? After all this searching and decoding you will still need to assemble the words into a sentence and extract some sort of meaning. To perform all these tasks in a reasonable period of time (i.e. today, rather than next week) is way beyond even the most powerful computers.

On the other hand, you could be using your knowledge to generate "expectations". Knowledge of the rules of English grammar will tell you, for example, that most correct English sentences are composed of a noun phrase and a verb phrase. Finding the verb phrase is (there it is!) the key to understanding the sentence. Even if you have not studied the formal rules of English grammar you probably are, and have been, using them to read. Unfortunately, grammar alone does not enable you to understand a sentence. Knowledge of the world seems to be an important element in being able to extract the meaning from a sentence. It can also tell you in advance what sorts of words and sentences might be coming next. Using world knowledge to reduce the amount of computation involved in reading solves one immediate problem, but at a high price. Now we need to find out how you get that world knowledge and how you can access it so quickly in order to be able to read this. Getting a machine to do this is not going to be easy!

Please don't become so self-conscious about your reading that you don't go any further! What is important here is that you don't need to ask these questions to be able to read, but you do need answers to most of them in order to build a reading machine or to give a scientific account of just how it is that you can read. AI makes us at least consider how we might begin the task of building a machine to do something, even if no such machine is actually in prospect. That, in turn, imposes far more intellectual and scientific rigour on the way we look at examples of intelligent behaviour. Even if the machine we consider is just a theoretical machine that we can't yet build, the discussion must get beyond superficial accounts of how we do things. Just as the crudest of aircraft ended the "power of flight" approach to bird flight, so consideration of even the crudest intelligent machines means that we must look at biology, psychology, and linguistics in a much more rigorous and detailed way.

## what is the ultimate goal of ai?

As we have seen, AI involves a tremendously wide range of problems and approaches. In fact this range is so wide that it often seems to some AI researchers that others in the field are not even doing the same subject. In practice this is not usually a problem. "Let a thousand flowers bloom" is a popular slogan in AI. For some commentators, on the other hand, it is important to be able to give some account of the "ultimate goal" of the subject. One research team, for example, might be spending their time machining precision gear-wheels with the goal of producing a robot that can walk up stairs without falling over. Another team might be analysing literature to see if they can determine any rules underlying the use of metaphor. Their ultimate goal might be a computer program that can recognize and respond to metaphors in human input. How can we say that they are, in some real sense, engaged in the same project?

Saying just what is the final goal that unites such different strands of research is not easy and wrong answers may have unfortunate consequences. Over the history of AI, there have been a number of attempts to make a single simple description of the ultimate goal of AI and they have generally been unsatisfactory. Many contemporary researchers would prefer not to confront the problem. They prefer instead to concentrate on their own local goals. However, in doing so they may be missing the sort of interdisciplinary crossover that has often proved so useful in AI. The gear-wheel machinists may need to know, for example, how insect legs are articulated, what birds need in order to be able to balance on two legs, and so on. The literature analysers may need to know about work in multi-valued logic which might bear on the ways in which metaphors work. In spite of all the difficulties, it is still worth looking at some of the suggested answers to the question "what is the ultimate goal of AI?"

We have already seen one possible approach and my personal favourite in the "aerodynamics of intelligence" mentioned in the preceding section. This approach would claim that the ultimate goal of AI is to produce a full scientific account of human, animal, and machine intelligence showing the common principles underlying all three. The problem with this, it must be admitted, is that we know very few, if any, of those common principles at the moment. We will look at this in more detail in chapter 5.

Many of the other approaches to defining the ultimate goal of AI have tended to stress the development of "human-like" levels of intelligence in machines. These need to be treated with caution for the reasons we have already seen. However, one of them – the so-called "Turing test" has had such a tremendous influence on the history of AI that we must spend the next two sections examining it in detail.

## the turing test

Undoubtedly the most famous answer to the question "what is the ultimate goal of AI?" is provided by the so-called Turing test. I say "so-called" because Alan Turing, after whom it is named, never talked of a test. Indeed, there are so many misinterpretations of Turing and the test attributed to him that it might accurately be called AI folklore. With your indulgence, I would like therefore to relate the whole story in some detail.

Alan Turing was undoubtedly a genius. Shortly after graduating in mathematics at Kings College, Cambridge he wrote a paper (published in 1936) which revolutionized our understanding of the nature of mathematics. That would have been enough for the average genius, but for Turing it was only the beginning. During World War II (in September 1939 to be precise) he and a number of high-powered intellectuals were secreted away by the British military in a requisitioned stately home called Bletchley Park. This is now part of a suburb of Milton Keynes in the South of England and is well worth a visit.

There they worked on breaking German military codes known as Enigma. In this objective they were supremely successful. Turing himself played a central role in understanding how Enigma was in fact breakable. That the code could be broken was never considered even possible by the Germans. Even at the end of the war, when it was obvious that the Allies had advance knowledge of German movements, the German high command was searching for traitors rather than considering the possibility that the Enigma code might have been cracked.

It should be obvious that the British ability to decode most German secret transmissions was a war-winning advantage. Even the most conservative historians admit that this achievement shortened the war by at least a year. Less obvious is the fact that it all remained highly secret. In fact, the truth about what had happened at Bletchley Park did not begin to emerge until the 1980s, and some aspects remain classified even now. Most importantly for present purposes, the code-breaking work at Bletchley Park involved the use of machines which were the precursors of modern computers. The Enigma code was so called because it was generated and decoded by the Enigma machine. Various other machines were used by the British code breakers. The most important of these was known as Colossus. This had most of the features of modern electronic computers, but in the rather foolish desire for total secrecy all ten of these machines at Bletchley Park were totally destroyed at the end of the war.

This left Turing and his colleagues in an embarrassing position. They knew then enough to build effective electronic computers, but they could not really say how they knew. The fact that they had seen such machines working day in and day out at Bletchley Park could never even be hinted at. Eventually a small team built a machine at Manchester University. It is from this machine that all modern computers are descended. In 1948 Alan Turing was writing programs for this machine. He was also writing a paper entitled "Computing Machinery and Intelligence". This paper laid out the ideas which became known as the Turing test.

"Computing Machinery and Intelligence" was published in 1950 in *Mind* – one of the longest established British philosophy journals. Let us note that it was a paper written by a mathematician, turned code-breaker, turned computer programmer, and was published in a philosophy journal. The interdisciplinary nature of AI has been apparent since its very outset.

In "Computing Machinery and Intelligence" Turing says he wishes to discuss the question, "Can machines think?" However, since this question is too vague, he proposes replacing it with a game. This game he called the "imitation game". It involves three people in separate rooms. They can communicate only by typing messages to each other. In the original version, there is a man, a woman, and an interrogator whose gender is unimportant. The interrogator, as the name suggests, can ask any question of the other two participants. The objective of the game is for both the man and the woman to convince the interrogator that they are the woman. The woman will be answering truthfully and the man will be typing things like "Don't listen to him, I'm the woman". (It's rather like what goes on in some Internet chatrooms.)

Now what would we say, asks Turing, if the role of the man in this game were to be successfully played by a machine? That is, if, after five minutes of questions, the average interrogator would not be able to recognize that he or she was communicating with a machine on at least thirty per cent of occasions. If we could make machines that could do this well in the imitation game, then ordinary people would be happy to say that they were thinking machines.

In fact, Turing thought that it was a matter of "when", not "if", we would make such machines. He confidently predicted that, by the year 2000, digital computers would be able to achieve this level of success in the imitation game. This achievement would change public attitudes so that it would become normal to talk of "thinking machines". One remarkable point about this paper is that Turing managed accurately to predict the level of computing power that would be available by the year 2000. This was despite the only real contemporary example being the Manchester machine whose roomful of equipment had much less computing power than we get from a tiny microchip nowadays. Turing was right in his prediction about the growth of computer power; however, no computer is anywhere near good enough to succeed in the imitation game in the foreseeable future.

Unfortunately, Alan Turing's career ended not long after the publication of this paper. He committed suicide in 1954, still only 42. One interesting final point about his life is that during the 1950s his interests and publications had moved on to the mathematical foundations of biology – an area that began to excite AI researchers again in the 1990s.

## is the turing test the ultimate goal of ai?

There are a number of reasons why the Turing test should not be seen as any sort of goal for AI, least of all the ultimate goal. Firstly, it concentrates on *human* performance and that is an unnecessary restriction for AI. AI is also concerned with other animals, most of which could not participate in the "imitation game". Secondly, in the actual building of machines, it is a distraction constantly to have to imitate human methods and performance.

Some people in AI do not agree with my last paragraph. Indeed, candidate programs are still entered into an imitation game competition every year. This is known as the "Loebner prize" after Dr Hugh Loebner, the inventor and industrialist who has offered a prize of 100,000 dollars for the first computer program to pass his version of the Turing test. Although no program has yet won the Grand Prize, there is a smaller prize of 2000 dollars for the most human-like computer program in the competition each year and this attracts a number of good attempts.

Looking at these programs in detail reveals another problem with treating the Turing test as the ultimate goal of AI. All of the 2000 dollar prize-winners have been fairly simple computer programs that are designed to *give the illusion* of holding a conversation. Nowadays such programs are called "chatbots". They have a number of set responses which they print out in response to various inputs by the interrogator. This is a technique which was first used in a program called ELIZA in 1966. The name is taken from Eliza Dolittle in Shaw's play

*Pygmalion.* It is not used entirely accurately, however, since Eliza Dolittle was taught to speak, while the eponymous program merely gives the illusion of being able to speak. There has been some refinement of such illusion programs since 1966, but these refinements do not really contribute to progress in AI. For example, if the program prints out text that is highly opinionated about matters that are political or sexual, then interrogators are more likely to think that it is human. This tells us something about human psychology which might be of some interest, but it tells us nothing about how to build intelligent machines.

This reveals a third serious problem with treating the Turing test as the ultimate goal of AI. It pushes researchers to produce programs that are primarily aimed at deceiving humans, not at any more fundamental approach to the problem of intelligence. In the next chapter we will look at some programs for which the claim of passing the Turing test has been made. It should quickly become obvious that they are actually more about deceit than about intelligence.

Many people working in AI would agree with my criticism of the Turing test, but say that it is still relevant because a truly intelligent machine (whatever that means) would be able to pass the test – mainly as a by-product of being intelligent. This may or may not turn out to be the case, but we are unlikely to see such a machine built in the near future. Remember that the interrogator may ask *absolutely any question.* This makes the Turing test a very hard test indeed.

Realistically, it is extremely difficult, expensive, and ultimately pointless to set about the project of building a machine that could pass the Turing test. Direct mimicry of human intelligent performance is unlikely to prove profitable when there is plenty of human intelligence available.

## further reading

An historical account of the early enthusiasm about AI in the US with many anecdotes and details of the personalities involved is provided in Pamela McCorduck's book, *Machines Who Think* (1979).

The best place to find out about Alan Turing and his achievements is in *Alan Turing: The Enigma of Intelligence* by Andrew Hodges (1983). Hodges also maintains a large and truly comprehensive Alan Turing website at:
http://www.turing.org.uk/turing/index.html

Turing's "Computing Machinery and Intelligence" (published originally in *Mind* in 1950) is an easy to read and non-technical paper. It has been published in many places – including a good collection of thought-provoking papers: *The Mind's I* (Hoffstadter and Dennett, 1981).

A must-read on the amazing history of Bletchley Park is *Britain's Best Kept Secret* by Ted Enver (1994).

You can read all about the Loebner prize online at:
http://www.loebner.net/Prizef/loebner-prize.html

A wonderful book which makes clear just how the same principles of aerodynamics apply to both birds and aircraft is *The Simple Science of Flight: from Insects to Jumbo Jets* by Henk Tennekes (1997).

It also helps to learn to fly. Everybody should try it.

# ai at work

## some glittering successes

When you watch the space shuttle climb slowly away from
Kennedy Space Port bound for another spell in orbit you might
well be impressed by the raw power required to escape from the
planet and the engineering skill required to harness this power.
Equally impressive, but less obvious, is the fact that the shuttle
can only fly as the result of an AI program. This program works
behind the scenes to schedule the operations involved in
preparing the space vehicle for launch.

There are between five to ten thousand distinct engineering
operations involved in turning around a space shuttle following
touch down and readying it for return to space. These operations
are interdependent in many complex ways, in some cases
requiring sequences of other operations to be completed before
starting. In other cases starting a sequence of operations, for
example recharging a fuel system, may prevent any other work
being carried out on that system. In order for the turn around of
the space vehicle to proceed as quickly as possible it is vital that
this complex pattern of operations be completed smoothly. Any
mistakes, repetitions, or unnecessary work could delay the next
launch by several months.

What is more, the sequence of engineering operations cannot be set in stone and blindly followed. Every mission requires a slightly different sequence of operations as payload, trajectory, time in orbit and so on vary. Each mission involves different patterns of wear, stress, and damage to the space vehicle. These patterns may well only be detected when systems are dismantled as part of the turn-around. The order of operations will have to be changed many times as problems are encountered. This is sometimes called a "constraint satisfaction problem" and is the sort of problem we encounter when changing one thing means changing countless others in a complex interlocking web. Preparing a school timetable is another constraint satisfaction problem. The difference with the space shuttle problem is that millions of dollars are involved, mistakes are unbelievably costly and, given that launches can only take place at certain times, it's important to find the quickest possible schedule.

For these reasons NASA chose to build an AI system which would calculate the best sequence in which to prepare the shuttle. It is, in many ways, typical of AI at work. This area of AI is known as scheduling and the type of reasoning used by the program is known as constraint-based reasoning. Essentially this is the task of finding a solution to a highly interconnected set of problems. It is an area (though by no means the only area) in which AI has produced spectacular successes. This program is typical of successful AI in other ways. The program was not trivial or easy to build – it cost NASA close to two million dollars and took three years to build. However, this large and expensive program has repaid its cost many times over. NASA estimates that it saves half a million to a million dollars per flight – mainly in reduced overtime working. The tasks performed by the program could, in principle, be performed by manual methods. However, according to NASA, such manual methods would take at least ten times as long and would be much more error-prone. Were it not for the AI system, NASA would be lucky to launch the shuttle once a year.

As far as NASA is concerned, this program is a great success, but it is only one among many. NASA has a continuing commitment to researching and applying AI. The applications of AI in space range from autonomous robots to systems that give advice to astronauts. It is important to stress that these are not experimental or fanciful systems but examples of reliable working technology – reliable enough to be built into spacecraft. The first AI program to actually go into space was DEVISER, a scheduling program which controlled the operation of Voyager I, launched in 1977. This is now the most distant man-made object in the Universe so it is fitting that, as a technological ambassador for humanity, it should contain AI. Current research projects include entirely autonomous spacecraft which will explore deep space on our behalf.[1]

AI is not only successful in the high-tech world of space flight. AI systems are also vital in business and finance. Just as in the example of the space shuttle, they tend to work behind the scenes, but the work they do is essential to the way we now do business. AI has also become an essential part of computer games – an industry which has grown larger than the film industry. AI has also been successful in actually playing some games. In 1997 an AI program defeated the world champion in a chess match. These are spectacular successes indeed and it is hard to think of any modern technology which has achieved so much. This chapter will look at some of these spectacular successes in detail.

## searching for solutions

If there is one concept above all others that is fundamental to understanding how AI works, it is search. In AI, "search" describes a technique of searching for the solution to a problem, rather than simply looking for lost car keys. (Even so, AI search may be able to help find your keys – so it's worth reading on.)

This is a very general set of techniques which has been used in almost every area of AI.

The reason these techniques are fundamental to the notion of artificial intelligence stems from the nature of computation. At the most basic level a computer is a thoroughly obedient moron. Technically we say that it executes an algorithm – a pattern of simple ordered steps. You may be able to see how mathematical calculations can be broken down into simple steps that could be performed by our thoroughly obedient moron. In fact almost all mathematical calculations can be expressed as algorithms. (We will discuss why not *all* mathematical calculation in a later chapter.) The quest for AI very often starts with filling in the missing pieces between an interesting real world problem – the sort you have to deal with – and an algorithm or set of simple sets that the computer can perform. The most important and general way in which these missing steps are filled in is by turning the real world problem into a search problem.

The way in which any general problem is turned into a search problem is to divide it into three elements. These three elements are: a start position; a set of transitions from one position to another; and a goal or solution position. It turns out that a surprisingly large proportion of real world problems can be described or re-described in terms of these three elements. Once we have managed to describe a problem in these three elements, the task of solving it has been made into an algorithmic problem. That is to say there is a relatively simple set of operations which can be performed by a computer. Essentially the program will keep going through the transitions until it finds the goal.

Like all good ideas, the idea underlying AI search is very simple. However, its simplicity hides tremendous effectiveness so it is worth looking at the process in more detail.

An example might help. Let's assume you are incensed by my enthusiasm for AI and decide to write me a letter about it. You have four pens in a pot on your desk and want to find one that works. What you will probably do is to take out a pen, try it, and discard it if it doesn't work. You then will probably try a second pen and so on until you find a pen that works.

In the jargon of AI search, the start state was no working pen, the transitions were to try pens one by one, and the goal state was a working pen with which to write a letter. Your search strategy of trying each pen in turn has some useful features. In particular, you noticed the need to discard pens that fail to work, to avoid trying the same pen more than once. It's important not to waste effort in search. This is because the size of the problem matters. In this case there were only four pens and it would not take too long to go through all four – an "exhaustive search" in AI jargon. However, if I had said that there were four hundred pens, rather than just four, and only one worked, then you would probably not want to bother with an exhaustive search strategy. In this case the total size of the problem – the "search space" in the jargon – is too large for exhaustive methods.

Exactly the same process applies to computers. Computers can run through the steps of an algorithm quickly, but they still have clear limits. A problem with using techniques such as these soon emerged and became known as "the combinatorial explosion". Like much AI jargon this sounds rather grander than it actually is. The idea behind this term is simply that, in many problems the number of possibilities does not increase in a smooth or linear way, but at a far faster rate.

The best way to illustrate this is with an old Indian folk tale. It tells how King Shirim offered his grand vizier, Sissa Ben Dahir, a magnificent reward for inventing the game of chess. He would give Sissa a gold piece for each of the 64 squares of the chess board. The vizier politely declined and asked instead for a single grain of wheat on the first square, two on the second, four on the third, and so on doubling each time. The king was amazed by the modesty of this request but was persuaded to comply.

The vizier had played a simple mathematical trick on the king. In fact a relatively simple calculation shows that he could never grant Sissa's request. The vizier had simply asked for $2^{64} - 1$ grains. That works out at 18,446,744,073,709,551,615 grains of wheat. This would be about four centuries worth of world wheat production at present-day rates. This is obviously impossible for even the most powerful king and Shirim had to give up. It's only a legend, but the moral is clear – never underestimate the power of a geometric progression. If Sissa had asked for two gold pieces per square that would have been an obvious doubling of his fee but the king could have paid 128 gold pieces without difficulty. By making the doubling into what mathematicians call a geometric series, the vizier made an impossible demand – even for something so cheap as a single grain of wheat.

It is the power of geometric series to reach impossibly large numbers that frustrates using simple computational techniques to solve many problems through exhaustive search. As with Sissa's fee, the numbers "explode" into impossibility even for the most powerful computers.

However that doesn't mean that AI researchers gave up. What made – and to a large extent, still makes – AI different from other types of computing was the introduction of heuristics. An heuristic is a rule of thumb, an educated guess or a clue as to the solution of a problem. With many problems it seems as if the mathematics of searching for a solution are impossible, but the real world is usually different from the world of pure mathematics. An heuristic is a way of putting elements of the real world back into a problem. In the world of mathematics numbers are equal and indistinguishable; in the real world there are usually patterns and clues which can help us in our searches. In our example, when you searched for a working pen you simply tried any pen at random. A random walk through a search space is usually considered unintelligent in AI. Heuristics can give guidance as to which way to proceed through a large search space. Another way of saying this would be that an intelligent approach (by either a computer or a human) would take into account any clues in the real world.

To see how this can be, lets go back to the four hundred pens. It's unlikely that you would bother to hold on to 399 pens that didn't work, so let's make the example slightly more realistic by saying that I have hidden a £20 note or a $50 bill or equivalent in one of four hundred pens. Your problem is to find it. An exhaustive search probably doesn't sound worthwhile at this point, but let's look at those patterns and clues that the real world can give. Mathematically, four hundred pens is just a number – a "featureless search space" in the jargon. I don't know where you might get four hundred pens from but (here in the real world) my stationery supplies outlet sells them in boxes of twenty. So I would have to buy twenty boxes. In order to hide the cash, I'd have to open one of the boxes, take out a pen, put the money in the pen and put it back. So you could look through a search space of twenty boxes for one that had a broken seal. In that box you could look through a search space of twenty pens for one that looked like it had been taken out or opened. This would probably get you to the money far more quickly than an exhaustive search.

The word "probably" in the last sentence is important. Heuristics are not a certain method of getting to the goal. I could have opened all the boxes just to confuse you, in which case the methods suggested would be little better than an exhaustive search. If I tipped all the pens into a pile, then the search space would become almost featureless. In spite of this fallibility heuristics usually serve to guide a real-world search.

A final word on search: it might seem that all this has little to do with intelligence, artificial or human. Many of the techniques developed in the field of AI search are now considered part of computing and information technology in general. They represent one of the first of a vast number of really useful "spin-offs" from AI. What needs to be kept in mind is that AI has patiently had to discover the steps that lie in between the automatic following of a program by a computer and intelligent performance in the real world. It seems that there may be many such steps, but search is certainly one of the most general and most basic. It's also important to be clear that AI is not here simply trading upon the power of the computer as an adding machine. The combinatorial explosion means that adding machines can't solve many problems, so we need to add a bit of intelligence. The addition of heuristics allows searches to be guided to some extent by *guesses* about the nature of the real world. It's only the beginning of the story of AI, but it is a very good starting point.

To see a clear example of how heuristically guided search can enable a computer to perform an intelligent task, let's return to the 1997 defeat of chess world grand master Gary Kasparov.

## computer champions

In the early days of AI it seemed to many people that if they could get a computer to play chess successfully then they would certainly have achieved an important goal in the search for artificial intelligence. Nowadays it is possible to buy computers that play chess very well indeed in almost every mall or high street. This has led some people to assume that it must be easy to get programs to play chess or that chess is a basically computational game. Neither of these assumptions is correct and they fail to do justice to the brilliance and perseverance of those AI pioneers who made the early breakthroughs in the field of computer game playing.

In fact chess is an extremely difficult game to play using computational techniques. Firstly, it turns out that chess is an extremely good example of the combinatorial explosion described in the last section. In the middle part of the average chess game the branching factor is about 36. That is to say that you have a choice of about 36 legal moves. Because these possibilities "branch out" from the current position, people in AI often call it a "search tree" – though it is an upside down tree with its branches expanding as you go lower. Your opponent can respond in about 36 ways to each of your moves. So if you want to consider your next move, you have a field of 1,296 moves to choose from. If you want to think about the move after that then the number of possible moves has risen to 1,679,616. So the number of moves to be considered continues to grow at an impossible rate even for the most powerful computer.

By calculation, it can be shown that there will never be a computer powerful enough to play chess by working out all the moves. When a game of chess starts there are about $10^{123}$ (that is 10 followed by 123 noughts) possible board positions in the game. This is an alarmingly large number, more than the number of electrons in the known universe. No conceivable computer could ever calculate all possible moves and choose between them. Far cleverer techniques than this are required.

There is another problem in using pure computational power to play chess which stems from the nature of the game itself. Arthur Samuel, an important AI pioneer in this field, named this the credit-assignment problem. The credit-assignment problem is the problem of deciding which moves, if any, are winning moves. At the end of a game of chess there is a list of moves which have been made – some good, some bad. Even if a program has won the game there is no way of saying which, if any, of these moves have led to the win. In other words there is no way of assigning the credit to any move over any other.

Samuel solved both these problems in an ingenious and effective way. Actually Samuel was working on the game of checkers (draughts) but the method he used still forms the basis of modern chess programs. He introduced the idea of a static evaluation function. This is yet another heuristic in that it enables the program to take a guess about the best sort of move to make in a given situation. The idea is that the program looks at the board positions available from the present position and comes up with an evaluation as to how good each of them looks. This is a *static* evaluation in that it does not ask whether or not this board position will lead to a win, merely how good it looks. A board position, for example, in which one's opponent has fewer pieces looks relatively good and one in which it is your own pieces that have been taken looks relatively bad.

What a chess playing program does in essence is to calculate a static evaluation function for as many possible board positions as it can in the time available and then chooses the move that leads to the best (or least bad) seeming board position. Of course, because of the combinatorial explosion the program can consider relatively few possible board positions so this is a highly fallible technique. Equally fallible is the static evaluation function which, like all heuristics, is only a guess and might be wrong.

Static evaluation functions can be set by the programmers, but usually this is only a starting point. What Samuel did was to let two versions of the program play against each other. One had a randomly modified static evaluation function and the other did not change. If the modified version won then that would be adopted in future, if the original version won then it would be retained. Nowadays various other techniques, such as GAs (genetic algorithms) described in the next chapter, can be used to refine static evaluation functions but Samuel's original method is still one of the most effective.

If you consider your own playing of board games such as chess, you might be thinking at this point that this is still a rather wasteful way of doing things. You probably don't examine all the moves available to you at a given point in the game. Many will be obviously stupid. Well, a way of enabling programs not to waste time considering stupid moves was also devised. This is known as alpha-beta pruning. The pruning here is a horticultural metaphor suggesting that unproductive branches are being lopped off the search tree. When the program looks at possible board positions there are two classes that can be discounted. The first class is those that have high static evaluation functions but are unobtainable because the opposing player will not let the program move so as to attain them. The second class is those that are disastrous. If either of these situations is detected early in the search, then the search can be stopped at that point. There is no point, for example, in expanding the possible board positions which follow from a disastrous board position. This move is not going to be made and any effort put in to exploring just how bad things might go on to become is obviously wasted effort. Similarly, effort used to explore board positions that our opponent is capable of preventing us from attaining is the computational equivalent of day dreaming. Effort put into exploring how good things could be "if only" is also wasted effort.

However, even with all these clever AI techniques, chess playing still involves a great deal of computational effort. Static evaluation functions may consider 64 features of a board position (sometimes even more). Each of these features has to be calculated for every one of the possible future board positions. The total static evaluation function must then be calculated and the results passed back up the search tree. Use of alpha-beta pruning vastly reduces the size of the search tree but this only delays the impact of the combinatorial explosion. It does not remove the problem. In order for chess playing programs to look more than a couple of moves ahead (and doing this vastly improves their success) they need to run on very powerful computers.

On the other hand it is very wrong to assume that just because these AI techniques have been understood for decades and can be programmed into a computer, that chess playing (by both computers and humans) does not require a great deal of intelligence. Let's be perfectly clear. The computer does not play chess by simply calculating winning moves. This is mathematically impossible. The static evaluation function is a *guess* as to which move to make and is a guess which the computer refines through actually playing the game.

What is remarkable is that the pioneers in building programs to play board games like chess initially got so much of the basic work right. When Deep Blue beat Gary Kasparov in 1997 there were celebrations for the Deep Blue team and at IBM – the company funding the work and whose nickname (the Big Blue) is honoured in the name of the machine. For most people working in AI it was not much of a cause for celebration. This was not because they weren't interested or didn't care but, rather, because they had known for many years that it was only a matter of time.

## the power of knowledge

In the medieval mystery play, *Everyman*, the eponymous hero is required to undertake the great journey of life. Although most companions fail him in various ways, Knowledge says, "Everyman, I will go with thee, and be thy guide,/ In thy most need to go by thy side". It may surprise readers to learn that, just as knowledge is a most valuable and reliable companion on life's journey for humans, so it is enormously useful for a computer. One of the most successful of all branches of AI is known as "knowledge-based systems". It may not be obvious how a computer system could be given, like Everyman, the benefit of knowledge, but it turned out not to be as difficult as it might at first seem.

While many researchers were concentrating on heuristics and search as a route to intelligence, another very different approach was found. This new approach originated primarily at Stanford University in California. The idea underlying this approach was just as elegant in its simplicity. You don't need to perform a massive search for the solution to a problem if you simply know the answer. This solution could be stored in the computer in the form of a statement rather like, "If the problem is P, then the answer is A". More formally these researchers argued that an intelligent system could be relatively poor at reasoning if it was relatively rich in knowledge. Its power would stem from its having up-to-date real world knowledge.

Of course the elegant simplicity of this approach needs quite a few additions to make it work in practice. If the computer is to have a recognizable amount of knowledge there's going to be quite a lot of these statements. There will also be problems about discovering just what is the right statement for each situation, but these problems have been solved. The "if-then" statements are known as production rules – or just productions, since the general idea of a rule is slightly misleading here. A collection of a few hundred or more of these rules is known as a knowledge base. The program will also contain an inference engine which is the part of the program that sifts through the knowledge base so as to find the right production rule for a particular situation. Of course, such a system will be performing a form of search and will also use heuristics wherever possible.

Another vitally important component of such systems is the ability to explain its reasoning process in full. This is not just for show. Because such systems are designed to give timely and useful advice about real world problems, they cannot do so with certainty. You may expect certainty from a pocket calculator, but not from a system performing medical or engineering diagnosis or giving tax advice. In real world problems there is nearly always an element of uncertainty and the more interesting the problem and the more complex the analysis of the problem, the greater that degree of uncertainty tends to become. What is more, it is the explanation facility that allows the system builders to refine the knowledge base. By asking the system to explain its conclusion, they can see what alterations, if any, need to be made to the knowledge base and inference engine.

The resulting program is usually called an "expert system". I prefer the more accurate and less misleading "knowledge-based system". However, "expert system" is the name which caught on and such systems really do contain human expertise. It is the interaction between the knowledge base and the inference engine that captures both the knowledge and the reasoning process of a human expert.

Many readers will feel that this is rather an extravagant claim. Human knowledge is a complex thing. It is often the product of experience, rather than explicit learning. It usually involves subtle judgements rather than blind following of rules. It is important to re-state that the sort of AI systems I am describing really do contain and use exactly this sort of knowledge. A knowledge-based system does not simply contain facts and deduce from them. It contains judgements which will probably be uncertain and the reasoning performed by such a system is not purely logical. Typically, but by no means exclusively, a knowledge-based system will perform some sort of diagnosis. In reasoning terms this involves looking at a set of symptoms and from them finding the most plausible explanation of the observed symptoms. This is known as abduction and it is not a purely logical task. There is no certainty about the conclusion, or even of the steps towards the conclusion.

Can such a system perform usefully in real-world situations? Most certainly it can. A pioneering knowledge-based system known as MYCIN was developed in the mid to late 1970s as a cooperative venture between the Department of Computer Science and the Medical School at Stanford University in California. MYCIN's knowledge base was of the diagnosis and treatment of infectious diseases of the blood. In 1979 formal studies showed that MYCIN's performance compared favourably with that of the human experts at Stanford.

The test involved looking at ten past cases of bacteraemia and meningitis. Eight experts rated the suggested therapies on these ten cases as either acceptable or unacceptable with a 1 or a 0. A perfect score would therefore be 80. The results are shown in the table below.

| Expert | Score |
|---|---|
| MYCIN | 52 |
| Faculty1 | 50 |
| Faculty2 | 48 |
| Infectious diseases fellow | 48 |
| Faculty3 | 46 |
| Faculty4 | 44 |
| Resident | 36 |
| Faculty5 | 34 |
| Medical student | 24 |

I have set out the full results because they serve to illustrate far more than just the superior performance of MYCIN. The first important point is that none of the scores is anywhere near a perfect score. This reflects the fact that medical diagnosis is highly uncertain. Out of interest, the actual therapies that these patients had received was scored by the experts at 46. In the real world we cannot expect perfect solutions. The second point to note is that, although MYCIN's score was the highest, it is only by a relatively small margin. Knowledge-based systems can be impressive, but they are not magic. The sorts of problems that medical experts and medical expert systems have to deal with do not have single definite solutions. However, being able to apply AI systems to such difficult real-world problems is a technological achievement of the greatest significance.

MYCIN was only an experimental expert system but the technology which it demonstrated found many practical and useful applications. A commercially successful example known as the Authorizer's Assistant, was built for American Express. The name of this expert system is important and shows that it is an example of the second-generation expert systems. That is to say that it was from the outset designed to assist, rather than replace the human authorizer. It came up with recommendations, not directives, based on a range of data about a customer. This included account information, spending patterns, and personal data. From examination of various databases it produced a recommendation on credit authorization for an individual customer. It could also prompt the authorizer with any other relevant facts about this customer which should be taken into account. This use of a knowledge-based system to help a human quickly take a decision based on data that may be stored in a number of electronic databases is rather more typical of the many useful knowledge-based systems at work today.

## is knowledge the key to intelligence?

Since we have known for some time how to capture and use this sort of knowledge in an AI system, you might wonder why we don't simply build bigger and bigger knowledge bases and thus solve all the problems of Artificial Intelligence. There are many reasons why this can't be done (although that hasn't stopped some very optimistic researchers from trying).

The most important reason is, perhaps surprisingly, a problem which AI systems turned out to share with humans. Just as knowledge is immensely useful, so it is costly and time consuming to acquire. In the case of knowledge-based systems the knowledge is usually obtained from human experts. Now experts typically use their knowledge, they don't usually have to express it in ways that are understandable to an outsider. Human expertise will usually involve hidden judgements – what people often call intuition – based on many years of real-world experience. All this will have to be found out somehow and explicitly programmed into the system.

You may recall that it is up-to-date real-world knowledge alone that gives the knowledge-based system its effectiveness. The only practical way to obtain this is from a human expert, and it is usually done in the form of a series of interviews with a specialist known as a knowledge engineer. The human expert's experience, judgements, and intuitions must be discovered and then made explicit so that they can be built into the system in the form of production rules. This is both lengthy and costly. It's lengthy because it is generally rather slow – a reasonable rate of progress would be about three useful units of knowledge per day. MYCIN took about twenty person-years to build. And it's costly because an expert with up-to-date knowledge of a useful field tends to be able to ask a good price for her time. Good knowledge engineers don't come cheap either. It's not surprising that Ed Feigenbaum (of the MYCIN project) christened this "the knowledge-acquisition bottleneck".

Many attempts have been made to break this bottleneck and some are still very active research areas which we will meet later in the book. A good number of them proved useful as spin-offs for computing in general. One such was the idea of "fast-prototyping". To speed up the knowledge acquisition process it is usual practice actually to build a knowledge-based system as early as possible and then to refine its knowledge base by trying it against real-world cases and correcting what the experts declare to be incorrect output. The actual program part of knowledge-based systems (the inference engine) was refined so that nowadays one can buy an "expert system shell" for not much more than the price of a word processor and you literally have to "just add knowledge" to build a useful system.

One of the most important spin-offs from attempts to break the knowledge-acquisition bottleneck was the development of a field of research known as "knowledge elicitation". This is the study of how to extract knowledge – usually but not always from human workers. Obviously this is the main part of the knowledge engineer's task and techniques for doing it reliably and quickly were researched and refined. It is also a very useful process outside the field of AI. These techniques have led to greater understanding of the role of knowledge in how humans perform tasks – particularly, though not exclusively, within commercial organizations. Modern talk about the "rise of the knowledge economy" owes much to this spin-off from AI.

Most of the research effort, however, went into trying to find ways in which the computer could learn for itself. Human experts rely on good teachers, but they also learn from experience. If programs could learn from experience then the knowledge-acquisition bottleneck might at least be eased. Machine learning is the name of this area and, like so many areas of AI, it experienced tremendous early enthusiasm followed by a realization of just how difficult the problems were. Many systems were built that could classify or generalize from a set of examples. A wide variety of so-called "pattern recognition" techniques led to yet another useful and profitable spin-off, described in the next section. However, it is fair to say that learning, in both humans and computers, turned out to be rather poorly understood and pattern recognition is only part of the problem. As we saw in the last chapter, there are many elements of our human intelligence that we do not understand in any scientific way. Learning is one of these. The knowledge-acquisition bottleneck remains unbroken to this day.

There are other limitations to knowledge-based systems which prevent them achieving general-purpose intelligence. Most importantly, the knowledge they contain is of a fairly narrow area. It is of space shuttle engineering schedules only or of infectious blood diseases only. It is not general knowledge, wisdom, or common sense. Many unsuccessful attempts have been made to capture general-purpose knowledge and this problem motivates many contemporary AI researchers who are employing techniques that are light years away from knowledge engineering. This is a theme that will be taken up in later chapters.

For the present, knowledge-based systems remain highly successful, given certain limitations: they operate within a narrow niche; they lack common sense; and above all, they require a substantial investment because useful knowledge is costly to obtain. However, those organizations, such as NASA, who have made the necessary commitment, have found that their usefulness repays the cost many times over. Our modern life depends on such systems giving advice to doctors, nurses, engineers, financial advisers, astronauts, scientists, and all manner of specialists. Knowledge-based systems have permeated almost all areas of modern life. Indeed this area of AI has been so successful that many people no longer associate it with AI. They simply see advice-giving as yet another thing that computers can do. That is, I suppose, the ultimate success for any branch of science.

## diamonds in the data

One of the most important spin-offs of the research into machine learning was a set of AI techniques which became known as "data mining". Behind the analogy in this name lies the fact that modern organizations have vast quantities of data. Most retailers will have an electronic record of every transaction made in their shops for as long as they care to keep it. Patient records, police records, tax data, and the like are vast collections of data and these days it is almost all held on computers. In addition there has grown up an enormous industry engaged in trading customer lists, mailing lists, and so on. Decision makers are not short of data, but the problem lies in extracting anything useful from it. A data mining program is one which can find the "diamonds" in this vast mine of data.

Data mining is achieved by a rich combination of AI technologies. Some resemble those we have already seen such as knowledge-based search and pattern-matching. Data mining also involves techniques that are more biologically based which we will meet in the next chapter. In commercial terms, data mining has proved very successful both for those designing the software and for those using it. The ability to extract quickly useful gems from the vast quantities of data involved in modern business has proved so important to commercial success that data mining software has been able to command very high prices. In these respects data mining is typical of successful AI.

Let's look in more detail at the way in which this technology emerged from AI research into machine learning. We have already seen how useful it is to put real world knowledge into an AI program and also how difficult it can be to get hold of such knowledge. One possible solution to this problem might be to devise ways of making computer programs that could learn for themselves. The area of AI research known as machine learning attracted a good deal of research effort.

Now, when somebody talks about "machine learning" it is tempting to translate this into human terms. We are, after all, most familiar with our own learning processes and this is how we tend to see the problem for the machine. Consideration of human learning, however, does not tell us much about the machine learning problem. Recall for a moment that what is wanted in a knowledge-based system is a set of production rules of the form: "in situation S do action A". The problem of machine learning is how to automatically generate a set of such rules which accurately reflect the real world without having to have humans deduce them and program them in by hand.

Getting an AI program to do this involves many difficult steps. One of the first (and most difficult) is to enable the program to recognize "situation S". Let's assume that "situation S" is a patient with a set of symptoms. (It could just as well be a board position in a game of chess, a step in a logic proof, a movement of share prices, or a step in preparing the space shuttle for launch.) What we have is a collection of observations, some relevant some not, on which the computer must base its diagnosis. It is important to recognize that this is not a problem like deduction or mathematical reasoning. Different people with the same illness will show different sets of symptoms. People may have more than one illness and some people may display symptoms without actually suffering from any illness. To decide whether or not a given person is actually suffering from a particular illness is a matter of deciding which is the most probable cause of the observed symptoms. The technical name for this reasoning process is abduction and it is probably closest to the notion of an expert opinion.

The way in which heuristic search works so well suggests that usually there are patterns to be discovered in the information that the real world presents to us. Recognizing these patterns is a key component of machine learning, as this would allow the program to draw generalizations from sets of symptoms. In fact machine learning was able to draw on a wealth of pattern matching techniques from various branches of AI and to add some of its own. It turned out that many of these techniques, although very effective, were highly domain specific. That is to say that what works well for one subject area does not work so well in another. A repeated theme of AI research is the way in which a technique that is effective for one group of problems does not cross over to a different group of problems.

One thing that emerged from machine learning research therefore, was a collection of very effective techniques which very often would allow a program to detect a pattern in a large set of examples. Some researchers realized that this was useful, and indeed saleable, in itself. They transformed these AI techniques into easy-to-use software packages. With a good user interface it was made easy to explore large amounts of data looking for patterns. Good data-mining software does not use just one pattern-recognition technique. It offers the user six or more methods of exploring the data which can be used singly or in combination.

The result of this development was a wonderful tool for businessmen and scientists. For example, a data-mining package developed by a small AI company in England, known as Clementine, enabled a large multinational toiletries manufacturer to reduce its product testing on animals by ninety-eight per cent. What the scientists had was a long list of chemical compounds, some of which had turned out in the past to have unpleasant effects. What they did not have was any rule or detailed knowledge that would allow them to predict whether a new compound would be safe or not. The only way to find out was to test it on animals.

This is just the sort of situation where data mining can help. Running all the compounds, harmful and non-harmful, through Clementine together with a new compound produces a prediction – a guess based on any patterns in all this information – as to whether the new compound will be harmful. Only if Clementine predicts that the new compound is safe, need it be tested before it is included in a shampoo or toothpaste. This is just one of a vast number of successful applications for Clementine including its use in police investigations.

Machine learning may not have found a solution to the knowledge acquisition problem as originally posed, but in data mining it found something just as wonderful and every bit as useful. This is a repeated theme in the success of AI. Like Columbus setting out to find a way to India but instead discovering America, AI researchers have not filled our lives with robot butlers. What they have discovered is much more useful.

Playing chess is just the publicity element for IBM's RS6000 computer series, of which Deep Blue is the most famous example. Other computers in this series are being employed on less spectacular, but perhaps more useful tasks. These include developing new drug therapies, financial analysis, and weather forecasting. Data mining is a new technology and we have probably not yet seen all the benefits that it has to offer humanity.

Real AI products, such as data mining, simply don't attract the same amount of TV time or column inches as the direct human imitation versions of AI. However, unlike the human imitation work, these products are of great practical benefit. Like NASA's use of AI, they are often quiet, hidden machinery, but behind the scenes they have become essential to the way we live nowadays.

## further reading

NASA has a tremendous commitment to all sorts of AI research. Details of many current AI projects can be found by exploring the web site: http://www-aig.jpl.nasa.gov

Further details of how search works as an AI technique are in Thorton and du Boulay's *Artificial Intelligence through Search* (1992).

A good general introduction to the technology of knowledge-based systems is P. Jackson's *Introduction to Expert Systems* (1990).

Most AI textbooks cover chess-playing and Samuel's work. My favourite is Luger and Stubblefield's *Artificial Intelligence Structures and Strategies for Complex Problem Solving* (1993).

There's a web-site which details many of the achievements of the Clementine data-mining program at: http://www.spss.com/spssbi/clementine/

## notes

1.  On 1 February 2003, shortly after this was written, Space Shuttle Columbia was lost on re-entry together with seven of her crew. This tragic loss does not affect my claims about the usefulness of AI in planning space shuttle launches. I have decided to keep to the original text in the hope and belief that the shuttle will soon fly again.

# ai and biology

## introduction

Although the methods described in the last chapter have achieved some spectacular successes, it has always been apparent that most of such AI systems are very different from any comparable systems in the natural world. Over the history of AI there have been many and various attempts to find inspiration in the biological basis of intelligence. These explorations of the way in which life has produced intelligence have produced results that are just as spectacular as those of the last chapter, but very different. In this chapter we will explore the fascinating and productive relationship between AI and biology.

This relationship is very much a two-way street: biology also takes inspiration and learns from AI research. So this chapter will also attempt to give some sense of how AI has enabled biologists to explore their subject matter in new and exciting ways.

It is worth reminding readers that the whole of AI is a jargon-rich area, and biologically inspired AI has generally chosen not to use the jargon of search and knowledge that we have already met. Instead, it has shamelessly plundered the biological sciences for jargon that sounds more closely related to the natural world. To make matters worse some of the concepts given biological sounding names are exactly the same concepts that we have already met, but described in different jargon.

## inspiration from brains

### neural nets

Artificial neural nets are a type of computer program directly inspired by what we know about how the brains of humans and similar animals work. It's important to be a bit pedantic about this. It would be wrong to say that artificial neural nets are "like brains" even though, superficially, they are much more like brains than ordinary computer programs. First of all, despite the tremendous progress in neuroscience over the last fifteen years, we do not fully understand how even a single neuron (brain cell) works. It's even harder for us to be totally clear about how the vast numbers of neurons in human and animal brains support any kind of thought. Second of all there are many differences between artificial neural nets and brains. One of the most important is that brains are immersed in a complex and changing mix of chemicals which constantly affects their performance. That's why I put "inspired by what we know" in the first sentence. A further bit of pedantry is the need to remind readers that words like training and learning should not be taken anthropocentrically. These terms are used extensively in this field and it would be a mistake to think that they have exactly the same meaning that they have when they are applied to human learning.

### how computers handle information

In order to understand why AI researchers should seek more direct inspiration from brains it is necessary to look in a little more detail at how brains and computers differ. Modern computers are almost exclusively digital electronic devices. The word "digital" means not that they count on their fingers as the literally minded might expect, but that they represent information as numbers. In fact the only numbers used are one and zero (computers have but one finger!) but in practice this is plenty. All numbers can be represented by long strings of ones and zeroes and anything we can measure can be represented by them too.

To understand this more clearly let's consider the digital storage of music. The music is a complex pattern of rising and falling tones, beats, and all sorts of sounds. However, if we measure the music signal precisely many times per second, then this whole complex pattern can be represented as nothing more than a series of ones and zeroes. Transfer this series on to a CD and put it in a player, and the full complexity of the original music can be reproduced. The CD player reads the stored pattern of ones and zeroes from the CD as a complete set of instructions that enable it to reproduce the music.

Modern computer programs are just the same as the music on the CD – they are a string of ones and zeroes representing instructions telling the computer precisely what to do. As with the music, converting this to more interesting items is a long and tedious process but, luckily for us, the hard work has already been done. The word-processor I am using, for example, is giving a numerical value to each key stroke, checking the numbers in various registers (like the one that represents how many characters there are in this line already). Usually, it will send another number to the part of the computer that handles what is displayed on the screen which causes it to put that character in the right place on the screen. One of the main reasons this usually all works so well is that electronics are very fast and the whole process takes only a tiny fraction of a second.

All the working of a digital computer involves such a pattern of ones and zeroes. It is represented in the form of yes or no switches. The data enters in the form of a string of ones and zeroes and is stored in memory in a bank of thousands or millions of switches which are set either on or off (these days such switches are very small indeed). It is processed by a device called a

CPU (Central Processing Unit) and all this does is to examine a string of ones and zeroes and set various other switches either off or on. When the computation is finished the output string of ones and zeroes is converted back to something more interesting such as pictures or words on the screen or music from the loudspeaker. Once again, since every minute detail of the operation must be dealt with in turn by the CPU, it has to be very fast. When a modern computer is described as having a two Gigaherz processor this means that the CPU has a clock speed of two thousand million operations every second. To put this another way, if the CPU performed one operation per second (like the ticking of a grandfather clock) it would take more than 63 years to do what it actually does in a second. More informally: the tedious stupidity of the way in which digital computers operate is compensated for by their tremendous speed.

## handling information more like a brain

We have known for a long time – for the whole history of AI – that brains are not like this at all. Firstly, a neuron (brain cell) is rather more like a self-contained adding machine than a simple switch. Each of your neurons has a large number of connections (typically between 5000 and 20,000) to other neurons. The way each neuron seems to operate is roughly that it adds up (sums) the total amount of activity on all these connections and when this exceeds some level (threshold) the neuron will generate an output signal (fire). This output signal is itself an input to many other connected neurons and forms part of the input activity which they sum in turn and which may (if it exceeds the threshold) cause them to fire in turn and so on.

Secondly, there is no equivalent of the CPU in brains. Rather than examining every one or zero in turn, their highly interconnected nature means that there are constant swirling patterns of neurons firing and influencing the many other neurons connected to them. What is actually going on in your brain at this physical level, therefore, is a series of complex (actually *very, very* complex) patterns of neurons firing. It is worth observing at this point that the human brain has something like one hundred billion neurons. Most of them have thousands of connections to other neurons. AI has to work with numbers that are so much smaller than this that comparisons may be somewhat premature.

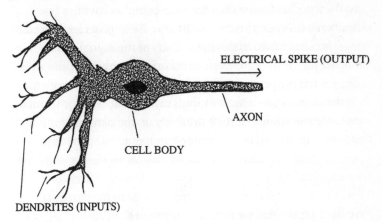

*A neuron*

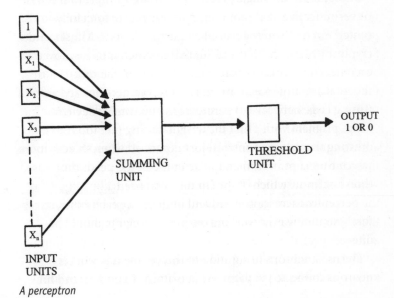

*A perceptron*

One early attempt to produce a computer that works more like a brain involved a device called a perceptron. A perceptron, shown on p.45, is a sort of electrical analogue of the crude picture of a neuron given above it. Technically, it is known as a threshold switching device. It has a number of input lines which it sums (or adds together) and a threshold value. If the result of the addition exceeds the threshold value then it fires (outputs one). If it is less than the threshold value then the perceptron will not fire (i.e. it outputs zero). A perceptron is relatively easy to program (the word "train" is usually used in this field). Each of the input lines is given a weight. The best way to think of these is as a volume control on each input. If the perceptron is firing when we do not want it to, then the weights on the active inputs can be reduced (the volume controls turned down) until it fires only on the pattern of inputs that we are interested in. If, on the other hand, it does not fire when we want it to, then the volume controls on the active inputs are turned up (the input weights increased) until it fires correctly. That is to say that it responds (by firing) only to the particular pattern of inputs that we want it to respond to.

Unfortunately, a single perceptron, though simple to train, is not very effective as a computer. This was rather forcefully pointed out in 1969 by a pair of AI gurus – Marvin Minsky and Seymour Papert – with the result that research interest moved away from this area for more than a decade. In the mid-1980s there was a resurgence of interest and researchers devised ways of training large networks of perceptrons and similar devices.

The problem with using the weight-setting method (or adjusting the volume controls) just described when there is more than one perceptron is that, unlike in the single perceptron case, we do not know which of the connections need adjusting. All of the perceptrons are connected and influencing each other, so there's no easy way of working out which weights should be adjusted.

The most important solution to this problem is what is known as the back-propagation algorithm. To understand how this works we have to consider a net that has three layers of threshold switching devices (perceptrons will do nicely). One layer is the input layer, then there is a hidden layer, then an output layer. Each threshold switching device (normally called a node) in the hidden layer is connected to each device in the input layer. Each node in the output layer is connected to each node in the hidden layer. The technical name for this is a feed-forward multi-layer neural network.

This type of net can be trained by passing a signal through from the input nodes, via the hidden nodes to the output nodes. Just as in the single perceptron case, training will consist of getting the outputs we want from these output nodes. The differences between the outputs we want and the outputs we get are represented as errors in the output nodes. These errors are then propagated back through the net so that the various active weights can be altered. Now we are fiddling with a large number of volume controls which interact with each other in a complex way. After a number of forward and back passes the weights are progressively adjusted so that the entire net behaves more or less

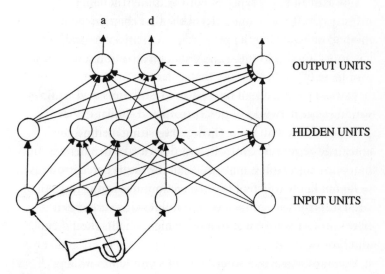

*Neural nets*

in the desired way. This back propagation algorithm is an effective, but rather slow, method of training artificial neural nets.

Please don't assume that this description precisely covers all artificial neural nets. There is an enormous number of different types of net used in AI. Many add to the principles just described by including varying degrees of randomness in their operation. This can be useful, for example in preventing the net moving too quickly to an immediate solution and helping it to relax into a more general solution. Others simplify the issue slightly by using nodes that are simple binary (i.e. on or off) switches, rather than threshold switching devices (switching at some point on a continuous scale). This does not represent a reversion to conventional computing. It represents a decision to rely solely on the highly interconnected nature of the network to generate its interesting behaviour.

Although the various types of net differ in many details and have different strengths and weaknesses, the basic principles of operation are as described above. Work is certainly progressing in this area and new types of networks are still being investigated. As was suggested in the first chapter, the best method of exploring the properties of artificial neural networks is to build them and try them against a set of problems.

However, there is one final point to add. *Almost all* artificial neural networks run as simulations on a conventional digital computer. This may seem a little perverse and confusing, especially since they represent a recognition of the fact that brains are very different from digital computers. It is important to remember that in this case the computer is acting just as a laboratory. As was said in the introduction, the speed and flexibility of the digital computer enables it to be used to build and investigate ideas like these. It is important to grasp that the items represented and experimented on do not themselves have to be computational items.

Nowadays meteorologists rely on large and complex computer simulations of weather systems in order to generate weather forecasts. Weather systems are certainly not computational in the sense we have seen. In fact they have much in common with brains. Just as in the brain there are constant swirling patterns of neurons firing, so the atmosphere is an ocean full of swirling patterns of temperature and pressure. Predicting these atmospheric patterns is very difficult; however, computer models help a great deal.

Artificial neural net research should be seen as basically similar. All the computer does is to provide a laboratory – a rich simulation which will enable some of the properties of such things as the weather to be captured with sufficient accuracy for forecasts to be made.

However, having introduced this analogy we need to understand its limitations. It is far too soon to call artificial neural nets simulations of the brain. Compared to even the simplest of animal brains they are small, crude, and one-dimensional. For example, we have so far talked of a neuron as rather like a self-contained adding machine. In fact a neuron may be rather more like a self-contained computer with the time intervals between firing functioning as a form of memory. Much more research is needed on biological neurons before we can consider accurately simulating them by these, or any other, methods.

There is simply no point at all in going on to compare artificial neural nets to the human brain. This has accurately been called the most complex object in the universe. The number of possible configurations of a human brain far exceeds the number of atoms in the entire universe. Forecasting weather patterns for a few days is very, very difficult. Forecasting the patterns in a human brain even for even a split second cannot realistically be contemplated with any foreseeable technology.

## what can artificial neural nets do?

It is one thing to grasp the idea of a neural net, quite another to understand how such a thing could ever act as some sort of a computer. Perhaps the most helpful metaphor here is drawn from physics. One could imagine the net as forming the surface of a table, which is distorted into a number of depressions by the learning process. A ball placed somewhere on the table would roll into the nearest such depression to its starting point. In other words, the entire net will have a number of states in which it is stable and when presented with a new input it will move into (the technical term is relax) the nearest stable state. When all this is working correctly this type of net will work as a sort of classification mechanism. It will be trained to classify a number of stereotypical examples. When presented with a fresh example it will relax into the state representing the stereotype which is closest to this particular new example. This classification process is one of the main applications of neural nets.

A task which illustrates this well is that of recognizing handwritten characters. The bar codes which nowadays seem to be attached to almost all products are easy for computers to understand. Actually they are just another form of digital signal, with the thick and thin lines representing a series of numbers and, in some cases, letters. (It's impossible to be more specific because, as is so often the case in modern computing, there are many different incompatible barcode conventions.)

Now humans don't often write barcodes, and even if they did they would not achieve the sort of uniformity that typifies mechanical methods. Human writing is nothing like so precise and predictable. Nonetheless, humans have been communicating by means of handwritten characters for centuries. Having learned to recognize a character – say one of the 26 letters of the Latin alphabet – we can then easily recognize slight variations on each of these letters. We certainly do not need the handwritten letter to be exactly the same as the first example we saw. We are very soon able to recognize even very malformed letters. Obviously there comes a point where the letter is so malformed that we cannot be sure whether it is, say, an "a" or a "d" but up to that point we have little problem reading handwriting.

Computers are not so adept at this. They usually need clear and consistent input – hence the use of barcodes. If we were to try to employ conventional computing methods to enable a computer to read handwriting we would have to define the shape of each letter as a set of points on a grid. The program would then have to calculate how close a scanned letter was to the ideal letter as defined by these points. Allowances would have to be made for variations in size and position. Long and complex calculations would have to be made by the program to cover all the different angles from which the character might be viewed. Even after this process, the program would need specific rules which explicitly define when the scanned character is more like an "a" than a "d" and so on for all the various possibilities. This is not just long-winded and difficult. It turned out not to work, either. The variation in handwritten characters is just too great for such precise methods to work. That's why there are barcodes and why computer-readable numbers have to be printed to a rigid format.

If we want computers to read variable and imprecise things like handwriting then artificial neural nets seem to be the way forward. We could, for example, train a neural net to recognize, say, the 26 lower-case character of the Latin alphabet. This training process would consist of presenting some typical or average examples of the 26 handwritten characters as inputs and adjusting the weights (volume controls) until we got the correct output. Since we want this net to classify the letters of the alphabet, a good design would be 26 outputs of which only one fires. After the training process it would have 26 stable states. When presented with a new character it would, if all goes well, relax into the state which corresponds to the stereotypical character closest to the new input. What is going on here is a form of classification in which we have not had to specify detailed rules of just what makes a given character more like an "a" than a "d". The rules for the classification are in the net in a sense, but certainly not in an explicit sense. We can't ever look at the net or the computer on which it is running and extract a rule for separating "a"s from "d"s. All we know is that the program is rather good at doing this. For this reason this approach to AI is sometimes called sub-symbolic.

## unsupervised learning

The example of training an artificial neural net to recognize handwritten characters described above is technically known as supervised learning. Supervised learning covers those cases where we know in advance what particular patterns we wish the net to learn and the training process involves matching the performance of the net to the recognition of those patterns. I have so far used examples where we present the net with a stereotype, or more usually a set of stereotypes. There is another class of applications for neural nets which is, in many ways, even more exciting. This is technically known as unsupervised learning and covers those cases where we don't know the pattern, or even if there is a pattern, and the task of the net is to find it.

This may seem a rather obscure use of sophisticated computing techniques, but in fact it is an area with many useful applications. One application for unsupervised learning that has already been described in the last chapter is data mining. Data-mining packages will offer their users at least one, often several, types of neural net. If a large and apparently unstructured data base is passed through the net then it may well be that patterns or clusters of data are detected.

There are some very interesting features of this approach to AI. First, there is the fact that it provides a method of classifying items that are messy and hard, if not impossible, to define accurately. It turns out that there are a lot of these messy items in the real world. The sounds of human speech are at least as imprecise as the shapes of human handwriting. Recognizing and interpreting the visual world through an eye or a camera may have much in common with classifying handwritten characters. It is certainly very difficult, if not impossible, to provide any hard and fast rules that will enable a computer to respond to visual input – to be able to distinguish a face from a flower pot, for example.

Second, because the sort of classification performed by a neural net does not depend on strict rules it is much more reliable when things get difficult. For many years AI researchers specifically sought a property called graceful degradation in their systems. The idea behind this is that most conventional computer programs either work perfectly or not at all. As most readers will have experienced with some frustration, when computer programs encounter a problem they tend to suddenly stop working altogether or go absolutely crazy. In these circumstances computer programmers say that a program has "fallen over" suggesting the rather accurate metaphor that it is like an overly tall stack of precariously balanced items which, once wobbled, will fall to the floor in a meaningless jumble. This description is remarkably close to what has actually happened.

This tendency to suddenly fall over is not so often seen in animal brains. Humans and other animals can certainly become confused or come to conclusions that are totally false. However, they do not suddenly fall over like computer programs. Even in the most challenging and confusing situations they usually attempt to do something. Developing programs that behave rather more like brains is a major area of AI research. This is what is meant by the quest for graceful degradation. Neural nets are one of the most promising ways of producing computers that degrade gracefully. Not only is this technically useful, but it also has important implications for our understanding of the ways in which animal brains might operate.

Because the whole neural net is involved in storing any particular piece of any information, it responds in some very interesting ways to any damage or deterioration. For example, if we were to train a net to recognize the alphabet in roughly the way just described, eventually we might be able to get reliable performance recognizing handwritten letters. If then we removed a couple of nodes from the net, it would not fail totally or fall over like a conventional computer program. Nor would it prove able to recognize only 24 out of the 26 characters. Instead its performance would degrade in a very different way. What would happen is that it would still classify all 26 characters, but in a slightly less reliable fashion. Perhaps its accuracy in making the correct classification might fall from about 99 per cent to about 80 per cent. Not only is this an example of graceful degradation, it is also quite similar to the deterioration in performance of a human or animal brain. (The experiment is often informally conducted by imbibing alcoholic drinks.)

Some researchers have produced results that look even more like the ways in which human brain performance can deteriorate. Artificial neural nets can be made to deteriorate in ways that are so similar to human neurological injury and illness that they are becoming a recognized research tool in human brain science and medicine.

## are artificial neural nets the key to intelligence?

If neural nets are so powerful, so interesting, and so like human intelligence, you might wonder why the route to solving all the problems of AI is not simply to build an enormous artificial neural net. Unfortunately, as with so many other approaches to AI, things turn out to be rather more difficult when we try to scale up or generalize a promising line of research. There are many serious problems with the idea of producing intelligence simply by building an enormous neural net. The successes we have considered in the previous section are (more or less) local solutions to local problems. Things get difficult if we attempt to go beyond this.

First of all, it is hard to train a net to perform many differing tasks. The ideal situation is where a relatively small net is trained to deal with a relatively small and, more importantly, well-defined problem. The nets we have considered so far are essentially single trick programs. The further we go from this, the more unpredictable and difficult the training operation becomes. It is most certainly not a case of, "today the alphabet, tomorrow the world". Ironically, perhaps, we have so far only been able to train artificial neural nets to deal with fairly well-defined problems of a manageable size. (We have, of course, been rather more successful with training our own natural neural nets.) It is not yet clear how an artificial neural net could be trained to deal with "the world" or any really open-ended sets of problems.

Now some readers may feel that this unpredictability is not a problem. After all, we are talking about training not programming and we expect a neural net to behave rather more like a brain than a computer. Given the usefulness of nets in unsupervised learning, it might seem therefore that we do not really need to worry about the problem being of manageable size and the training process being predictable. This is not the case; we really do need a manageable and well-defined problem for the training process to work. A famous AI urban myth may help to make this clearer.

The story goes something like this. A research team was training a neural net to recognize pictures containing tanks. (I'll leave you to guess why it was tanks and not tea-cups.) To do this they showed it two training sets of photographs. One set of pictures contained at least one tank somewhere in the scene, the other set contained no tanks. The net had to be trained to discriminate between the two sets of photographs. Eventually, after all that back-propagation stuff, it correctly gave the output "tank" when there was a tank in the picture and "no tank" when there wasn't. Even if, say, only a little bit of the gun was peeping out from behind a sand dune it said "tank". Then they presented a picture where no part of the tank was visible – it was actually completely hidden behind a sand dune – and the program said "tank".

Now when this sort of thing happens research labs tend to split along age-based lines. The young hairs say "Great! We're in line for the Nobel Prize!" and the old heads say "Something's gone wrong". Unfortunately, the old heads are usually right – as they were in this case.

What had happened was that the photographs containing tanks had been taken in the morning while the army played tanks on the range. After lunch the photographer had gone back and taken pictures from the same angles of the empty range. So the net had identified the most reliable single feature which enabled it to classify the two sets of photos, namely the angle of the shadows. "AM = tank, PM = no tank". This was an extremely effective way of classifying the two sets of photographs in the training set. What it most certainly was *not* was a program that recognizes tanks. The great advantage of neural nets is that they find their own classification criteria. The great problem is that it may not be the one you want!

Secondly, there are the problems of scale. The human brain is large, impossibly large as nets go, with one hundred to two hundred billion neurons. Nets anywhere approaching even a thousandth of this size are way beyond the state of the art at present. It takes very powerful computers and some simplifying assumptions to even approach the number of neurons in the brains of the simplest of creatures. A particular favourite in AI is *Aplysia californica*, the Californian sea-slug which has about twenty to forty thousand neurons, and this is still a very big net indeed in AI terms.

It is not just the difficulty of building (or simulating) such a large net that causes problems. The problems of training large nets do not increase simply in proportion to their size. The back-propagation algorithm, described above, is effective in training but very slow, and the more nodes there are and the more interconnected they are, the slower it becomes. These difficulties, in turn, interact with the need for a manageable and predictable problem on which to train the net.

An even more difficult problem of scale is the need to match the size of net to the size of the task which it is trained to perform. This problem is not as obvious as the ones just mentioned, but quite simple to grasp. If we trained a net to recognize, say, ten inputs, and the net had, say, ten hidden nodes, then each hidden node could act like a bit of conventional memory in a digital computer. Without noise or variation in the input then what would probably occur is something equivalent to conventional computing. The net would function simply as a look-up table and would not do any generalization. In short, it would behave more or less like a conventional program in a digital computer and would have none of the interesting properties described in the last section. There would be no guarantee that the net would use a distributed representation. This problem is even more serious for a very large net trained on a very simple problem.

These various problems of size and scale may yet turn out to be solvable. Work is continuing on these and related problems, but for the moment they effectively prevent any simple scaling up of the initial successes of artificial neural nets. Inspiration from the brain may suggest that instead of one enormous net, a number of smaller ones may be appropriate. The present state of neurophysiological knowledge, for example, suggests that the human brain contains about a hundred more or less distinct but interacting networks.

What artificial neural nets offer us is a tantalizing glimpse of how a part of biological intelligence works. There may well be more to come. This tantalizing glimpse has not been lost on neuroscientists and biologists. Artificial neural nets are a wonderful modelling tool and over the next few decades will very probably yield some interesting theories about how living brains operate. Once again we have the irony that AI has not been terribly successful in achieving some of its own declared goals, but has produced highly informative spin-offs for other areas of science. The last chapter will cover this in more detail.

## learning from evolution – genetic algorithms

Evolutionary computing is a field of AI that takes the basic principles of evolution and applies them in the form of a computer program. Instead of the programmer having to put a solution to a problem into the program, the program literally evolves a solution. To understand how this could work let's look in a little more detail at what I have so far simply referred to as the basic principles of evolution.

Modern genetic biology has shown that evolution takes place only through the selection and recombination of genes. These genes are analogous to computer code in that they represent essential information on the nature of a particular biological organism. Reproduction, and particularly sexual reproduction, allows the combination of successful genes. Evolution is not always well-described in scientific literature. However, I can be certain that my ancestors – and those of the squirrel, the trees, and the grass which I can see through my window – managed to survive long enough to reproduce themselves. All of us owe our genetic make-up to those ancestors.

The right sort of computer program can effectively simulate all this. The genes can be represented by bits of computer code (that's almost exactly what they are). As we shall see, the selection and reproduction processes can be simulated in the computer too. It is really not too extravagant to talk of such programs evolving solutions to problems because that is exactly what they are doing. However, the jargon warning at the beginning of the chapter should be recalled at this point, because all these computational notions are given very biological names.

A genetic algorithm (GA) is one particular form of evolutionary computing which we will consider in detail here. A GA acts in a largely similar fashion to biological evolution. The essential ingredient which can turn any problem into one which can be handled by a GA is the so-called fitness function. If we can describe a problem, any problem, in terms of some measure of how good any solution is then it should, in principle, be possible to evolve a solution to that problem. In fact this is another heuristic of the sort we saw in the last chapter. It is a feature (or set of features) of the problem that the programmer can use to guide the evolutionary process. In computing terms this is simply a method which will allow us to rank all potential solutions from hopeless to perfect. If such a ranking is possible then GA methods will, at least in principle, be able to evolve a solution to the problem.

What the computer does is to set up a set of candidate solutions to the problem (known as a population). At this initial stage they will be totally random. (To make the maths simple let's suppose that we have a population size of one hundred.) These hundred candidate solutions are then ranked by means of the fitness function. Even though they were generated as completely random attempts, it is possible to rank them from first to one hundredth. Of course, the chances are that even the first in this ranking will not look like any sort of solution to the problem.

The program now follows natural selection by dispensing with, say, the bottom fifty members of the population. They are replaced by combinations of the top fifty members of the population in a process inspired by biological reproduction. Portions of pairs of the top fifty are combined in a process known as crossover. This is the computational equivalent of biological reproduction. With the population up to one hundred again, the process of selection by means of the fitness function is repeated. Then the crossover is repeated, and so on for many generations. Eventually the program will converge towards an acceptable

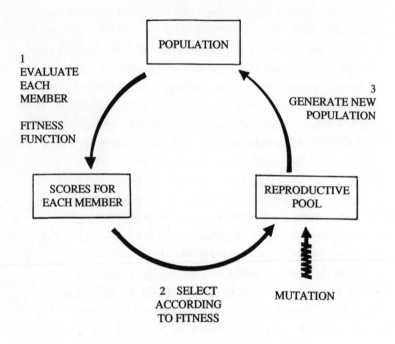

*Steps 1, 2, and 3 in a genetic algorithm*

solution to the problem – always assuming that there is at least one solution and that the fitness function can guide the selection process towards it. It's hard to say how many generations are required – this all depends on the size and nature of the problem – but for fairly simple problems it may only be a few hundred.

## a (literary) example

More cynical readers may feel that GAs have little to do with evolution and rely mainly on the ability of computers to calculate very quickly. Just as with search, chess playing, and expert knowledge this is most certainly not the case. There's more to this than number crunching. A detailed example may make this clearer.

It has been said that a sufficiently large number of monkeys sitting at a sufficiently large number of typewriters will eventually type the entire works of Shakespeare. I fear that the numbers involved will be too large for my ability to calculate so instead I want to work with just one short phrase: "methinks it is like a weasel". We can do a pretty exact calculation of just how many random key strokes it would take to produce this. There are 28 characters in the phrase (including spaces) and if we assume only lower case letters there are 27 keyboard characters (including the space bar) to choose from. If the monkey or computer just hit the keyboard at random then the chances of achieving this phrase would be: $(\frac{1}{27})^{28}$ which works out as approximately one in $10^{40}$ or worse than one in ten thousand, billion, billion, billion. These are very poor odds indeed and we can safely assume that it will never happen. It is possible, however, for a GA to vastly shorten the odds and produce this phrase by natural selection in a few seconds.

How would we write a program to evolve this phrase? Well the GA would start from a random position and, as we have seen, the chances of this being the correct phrase are so tiny that we can disregard it. However, we would add a fitness function. A suitable fitness function for this task would be a number which sums the closeness of each letter to the desired letter. Remember that a fitness function must allow the computer to rank all candidate phrases from totally random to correct. If the computer counts how far each letter is from the letter (or space) that should be in that position and sums them, then the phrase with the lowest score is the best (or the least hopeless).

Evolution is then imitated further by the program. Let's assume a population size of 100 phrases. The program ranks them according to the fitness function and dispenses with the bottom fifty. The top fifty can then be combined in a process that imitates reproduction. It's easy in this case to just take any pair of the top fifty phrases, break them at one point (say, the 14th character) and then combine the first half of one with the second half of another (and vice versa for the other two bits). We then have two new phrases (or offspring) and when this process is completed we will have a population of one hundred again. Most GAs add a very small amount of random change to stop things settling into purely local solutions. In this case one letter change at random per generation would be quite sufficient. The process is then repeated and steadily progresses towards the correct phrase.

If you think that a GA is just another search program – of the sort we saw in chapter 2 but given a lot of biological-sounding names – then you would be right. GAs are very powerful search programs because they consider a number of candidate solutions in parallel and they do not dispense with partial solutions.

Obviously the ability to run a simulation of evolution is an amazingly powerful research tool in biology. It enables biologists to answer questions about how creatures have changed over evolutionary time, how populations have changed, and about how they might change in the future.

Far from being an esoteric and theoretical novelty, evolutionary computing has proved very useful in engineering and commercial applications. One of the most powerful features of GAs is that they start from random. This means that they can evolve truly novel solutions whereas human engineers inevitably tend to look at what has been done before. Another benefit that they can bring is that evolved solutions tend to be robust. That is to say that because they have been progressively honed against a problem, they are usually not as brittle as something produced by a hands-on design process. When a solution to an engineering problem is found by means of a GA it usually tends to carry on working if things change.

Some of the areas that have successfully used GA-based techniques include the design of wings for airliners; the design of new financial services; and the design of electronic circuits. GAs are also an effective tool in data mining and for the pattern recognition applications discussed in the last chapter. As with most of the other AI techniques we have examined so far, the real benefits tend to be felt in fields other than pure AI.

## why not just evolve intelligence?

Once again readers may want to ask why it is – if GAs are so interesting and so easy to program – that they are not the short cut to producing artificial intelligence for which everybody is looking? The seductiveness of evolutionary computing is obvious. We know that evolution *has* produced intelligence. Indeed it is the only thing that has ever produced robust general-purpose intelligence. Why can't we just let our GA run on a big computer so that it will happen again? Again the answer lies in the fact that this branch of AI can produce impressive local successes but no general solution.

In fact this seductiveness of GAs is misleading. Although evolution has produced general-purpose intelligence, that is not what it is about or for. In fact, evolutionary processes are constantly trying to avoid producing general-purpose intelligence. A local single-purpose fix is almost always likely to be preferred. This stems from the nature of evolution itself.

Recall the fine examples of evolutionary success outside my window. Among them there's the squirrel, the silver birch, the grass, and plenty of dandelions at this time of year. However, only the squirrel has much of what we usually call intelligence. We would be very wrong to think that the added intellectual complexity of the squirrel makes it more successful in evolutionary terms. If conditions were to change – let's say global warming vastly reduced the rainfall around here – then the squirrel would no more be able to adapt than the tree or the grass. Indeed the dandelion's ability to preserve most of itself underground would render it the most successful. Of course the dandelions would face stiff competition from new drought adapted plants moving in. Indeed, the need of the academics for large quantities of coffee and therefore water might mean that the university too would have to be abandoned to be colonized by different (almost certainly non-intellectual) organisms. In short, biological evolution has no measure of *how* you survive, just *if* you survive. Intelligence is not usually required. There are also some good reasons for believing that *human* intelligence does not contribute to survival at all. We will consider these in detail in the next chapter.

A further problem is that of programming beyond the single-problem case. The researcher programming the GA may have a vague idea what she or he wants to evolve. The program, however, will evolve only what the fitness function directs it towards, if that. In other words, GAs aren't as effective as they might seem in very open-ended situations. We don't yet know enough (in scientific terms) about general-purpose intelligence to be able to provide a clear fitness function for it. That's not to say that there is no future in using evolutionary computing in AI: undoubtedly it is a fascinating part of current research and may well produce many more useful applications. Unfortunately it is no short cut to building an intelligent machine.

Just as with other successful AI techniques the fact that they are no short cuts does not diminish their usefulness. Crossing the Atlantic was no short cut to India (though that didn't stop people trying to find one). Evolutionary computing and artificial neural nets are relatively new technologies which have found many practical applications. They probably have much more to offer.

Biologically inspired AI has produced a wide variety of useful working programs in some rather diverse and unlikely areas. Even there they are usually behind the scenes. Just as with the space shuttle scheduling program, we tend to see only the airliner. It's much harder to see that its wing shape has been evolved by a GA, rather than designed by a human. Biologically inspired AI, just like the techniques described in the last chapter, is tending not to get the full credit for its successes. It's hard to tell that the latest generation of washing machines use artificial neural nets to control the wash programmes, but they do. We may not yet be able to build something approaching an artificial brain, but so much has been discovered in the attempt. There can be few, if any, areas of science that have produced so many useful spin-offs.

## further reading

Artificial neural nets are covered in general AI textbooks. A specific and readable introduction to them is provided in *An Introduction to Neural Computing* (Aleksander and Morton, 1990).

A good explanation of why there is much more to evolution than in popular accounts can be found in *Darwin's Dangerous Idea* (1995).

A good textbook which details these biologically inspired approaches to AI is *Understanding Intelligence* (Pfeifer and Scheier, 1999).

# some challenges

Having looked at what AI has achieved, in both practical and theoretical terms, it is time for a quick tour of what the main possibilities are at present. Before going on to the possibilities, however, we need to make a brief detour into some of the big challenges facing AI research. This detour will help explain some of the reasons why people outside the field have criticized AI and even why those in AI are often so critical of each other.

To set this in context it is worth remembering that AI represents one of the biggest (probably the absolute biggest) scientific challenges that humanity has ever undertaken. I said in chapter 1 that it was more difficult to achieve than space travel, and that is no exaggeration. It would be naive to assume that such an ambitious quest would not have to face many difficulties and criticisms. AI has certainly had its share but that does not mean that progress has been slowed or that enthusiasm has been lost.

It is also worth remembering that, unlike space travel, AI research does not require impossibly expensive equipment or place restrictions on those who may participate. The main qualification required to enter AI research is that of being able to think creatively but persistently about some very hard problems. For that reason, AI research is carried out in many countries and by many different people with very different backgrounds. When

I talk, in the rest of this chapter, about what people in AI often call "problems," readers should see them as puzzles that they themselves might be able to solve.

Many of the criticisms of AI over its history have come from outside the field. Philosophers in particular have frequently claimed that all or part of AI is misguided or impossible or both. We'll look at the more famous of these later in this section. However, there are also some practical problems which need to be examined before going on to the more philosophical ones. The philosophy with which we shall be concerned here is not particularly technical – it is just that the disagreements I wish to discuss are (as disagreements so often are) about ideas.

## factionalism

As a matter of history, AI researchers have tended to form into opposing camps and, on occasion, hurl abuse at each other. Now this seems to be a natural human tendency – what we might call a "known human defect" and is one of the reasons for people to become interested in AI in the first place. Ordinarily there would be no reason to remark on it. In AI, however, it has been, and still remains, an obstacle to progress.

Recall the military metaphor in chapter 1. AI researchers are pursuing their subject across just about the whole front of human knowledge. People have developed and continue to develop programs, and robots of every type to do just about everything that humans do. Writing poetry? Certainly, though I have never yet been much inspired by poems from any non-human. Exploring Mars? Yes, that looks like being a robot-led activity for some decades. Researching the effects of brain injury through neural simulations? That too is an important application for AI techniques.

Not only is there this broad spread of problem areas for AI, but there is almost as wide a spread of techniques employed. We have seen how heuristically guided search provides a good basis for certain types of problem solving which has led to world class game-playing programs. We have also seen how programs inspired by brains can solve problems (usually slightly different sorts of problems) in very different ways. Other researchers approach the area by building real physical robots. This diversity of techniques would not be so bad if only they could be combined, but AI researchers all too often tend to have only one of these techniques in their tool bag.

What is worse, they very often have nothing but contempt for the other techniques. They may see researchers using other techniques as fundamentally mistaken – or as not pursuing the same goal. There are many practical reasons why the short history of AI has thrown up this problem of factionalism. Researchers need to compete very hard for funding. In this competitive environment it seems to help loudly to distinguish one's own approach from those of competitors.

Also a large share of the really interesting work has been done by bright academic researchers at the start of their careers. Very much an infant science, AI is the sort of area in which an individual with creativity and ability can develop a novel idea into a working prototype more or less single-handed. This has led to many promising starts but it is not a recipe for combinations of techniques. However the problems of competing for funds and short-term small-scale research projects are not unique to AI. Biology, physics, and chemistry all share these problems to a greater or lesser degree.

What is special about AI is that AI researchers have worsened the problems of factionalism by over-generalizing their results and their claims. This has happened and continues to happen for all the approaches we have discussed in the previous two chapters. The researchers at the forefront of search-based methods claimed that this alone was the basis of intelligent behaviour in humans, animals, and machines. So did the advocates of knowledge-based techniques. So did the proselytizers for neural nets, for genetic algorithms, and for behaviour-based robotics. It is, one must suppose, just too tempting on discovering a successful technique to assume (or at least to claim) that this is the one and only key to mysteries of intelligence.

Intriguingly, this factionalism seems to matter much more in academia than it does in the business world. Here integration does seem possible. Many successful commercial AI products involve combinations of techniques. Data mining would be a good example. Here search-based, biologically inspired, and statistical analysis will all work together in a single program. The user can combine a neural net with a heuristically guided search simply by linking the icons on the desk top. In the commercial world the choice of technique seems more influenced by the question, "does it work?" rather than, "is it consonant with our particular orthodoxy?"

Another problem is that successful AI often involves very large projects – NASA's Space Shuttle scheduling program, described in chapter 2, would be a good example. This took three years of intensive effort to build and cost between 1.5 and 2 million dollars. Again, the need to build very large systems is not unique to AI. Computing in general is at the stage of development where large projects are often the order of the day. However, it is not easy to manage a large computing project. Industry experts estimate that, at present, about fifty per cent of all Information Technology projects simply fail to be completed. This is obviously a serious problem for AI research but it is one that is hardly ever recognized in the literature.

In summary, AI has to deal with all the problems that face large computing projects – which we still need to learn how to organize successfully. It also has to deal with the problems of science in general – such as bright but individualistic researchers in opposing camps. It has to deal with criticism, often hostile, from outside the discipline. In spite of this there has been steady progress. The next sections go on to describe some theoretical puzzles that are distinct to AI.

## what does the robot need to know?

If we are to provide general-purpose intelligence to enable a robot (or some other sort of machine) to be able to perform independently in an environment, then we need to be able to provide some ways in which the robot can take in information about that environment. However, this simple idea leads straight into some of the biggest challenges in AI. It is a simple matter to connect a television camera to a computer but it has proved extremely difficult to get the computer to extract any useful information from the signal from the camera. The robot could, of course, explore its environment through touch, radar, sonar, or some other form of sensing but the problems we are considering in this section would be just the same.

In fact there is a whole range of problems involved in achieving computer vision. However, we may characterize one fundamental problem in this area as the object-recognition problem. It has close links to a family of similar problems in other areas of AI. It would be useful, for example, if the robot's vision system could enable it to tell the difference between a cell phone and a glasses case. If I asked the robot to pass me the phone, I would like to think that on most occasions it passed me a phone rather than a glasses case.

This might seem a trivial problem to solve, but it has proved remarkably difficult. The most obvious approach to solving this problem is to provide the robot with knowledge of what a phone looks like and what glasses cases look like. This approach spirals into impossibility in the real world because what these objects look like depends on the lighting, the angle they are viewed from, whether or not there is anything else close to them, and so on. The problem simply multiplies if we consider different types of phones and different colours of glasses cases. In order to make some headway with this problem AI researchers usually resort to taking it out of the real world. In a highly predictable controlled environment then progress can be made with this group of problems. This is why robots can be usefully employed on production lines.

The problem then becomes how to enable the robot to deal with richer, less predictable situations. Given that we already know that knowledge is very powerful in solving real-world problems, a natural assumption would be to build plenty of real-world knowledge into the robot. Surprisingly, perhaps, this approach runs into difficulties. One class of difficulties with this approach has come to be called the "frame problem". It is more of a problem for knowledge-based approaches to AI but other approaches may have to face this problem, or something rather like it, eventually.

In fact, there are two distinct types of frame problem and a family of related problems but they all relate to the question, "how can we ensure that what the robot knows corresponds to the world – especially if the world changes?" The "frame" part of the name refers to the frames in a cartoon animation. From one frame to the next almost everything remains constant. One tiny feature – say, the position of Mickey's arm may change and it is this which provides the illusion of motion when the animation is viewed. This is a problem for AI because it is clear that we (and presumably any other intelligent entities) need to focus our attention on only a very small part of our environment – the bit that matters – but we also need to pay attention to side effects.

How is this a problem for AI? Well, consider an intelligent robot that has to deal with a real environment – say, your living room. In order to find its way around without bumping into anything, it needs to know where the walls and furniture are. If it goes out of the room and comes back in it can assume that the walls are in the same place but the furniture may have been moved by someone. If the robot is not to bump into things it needs to know what things can move and look out for them somehow.

On the other hand, there are many things about your living room that the robot won't ever need to know. You may have used the room for years without ever calculating its exact volume or its alignment with respect to magnetic north. One general way of describing this problem might therefore be, "How can we decide which features of your living room a robot would need to know and which features it would not?"

At best, this problem is a combinatorial explosion of the sort we saw in chapter 2. The list of things that might be known about your living room seems to go on indefinitely. Does the robot need to know the colour of the walls? Maybe it does, but how about the colour they used to be before you re-decorated? What then about whether or not it is bigger than the bedroom? Whether or not it is bigger or smaller than the Jones' living room?

You may have thought of an obvious solution to this problem. It might very well be something along the lines of what the philosopher, Andy Clark, calls "the 007 Principle" (presumably in reference to the world of international espionage) – "Know only what you need to know to get the job done". This is a good slogan for the robot in your living room but it does not solve the problem. The problem of working out what you need to know and what you don't need to know to get the job done is just a restatement of the original problem. If we programmed our robot with the 007 principle and set it to clean your living room, it would remain motionless while it decided that it did not need to know the previous colours of the walls but it did need to know the position of the table and so on. Since the list seems to be endless you would be well advised to forget the robot and clean the room yourself.

The fact that you *can* clean the room yourself suggests that such problems are not insoluble and may be more of an amusing paradox than a real difficulty. However, it may also be an important reason why, as we have seen, AI successes are hard to scale up so that we can build general intelligence into a computer or robot.

For a constrained world with a known set of problems we may not have to confront this problem. If we wanted to build a robot to work on a production line for example, then we can assume that there will only be a limited number of things that can happen. This number could be quite large in human terms since a computer can consider thousands of possibilities in a second. What is important is that it would be *limited*. If something strange happens then it can be programmed to call for help. In this case the number of things that it needs to know will be manageable. Unfortunately this may not be quite what we mean by "general-purpose intelligence". In the real world, if that is what we can call your living room, the number of things that can happen is far larger and may well be totally unmanageable. Later in this chapter, we shall see why some AI researchers think they may have found ways of avoiding this problem.

## trapped in the chinese room?

One of the most famous (though that does mean one of the best) criticisms of AI methodology is known as "The Chinese Room". This is the name of a thought experiment proposed by the philosopher, John Searle, at Berkeley. Searle does not object to the very idea of intelligent machines, indeed he believes that we humans are just that – thinking machines. What he claims is that a machine cannot become conscious simply by enacting a computer program. "Consciousness" is very difficult to define and is the source of endless debate. Luckily for present purposes Searle's objection does not depend on such slippery concepts, it depends more on the nature of computer programs.

Searle bases his objection on the fact that computer programs are basically algorithms. An algorithm is a set of steps which completely describe how to carry out an operation. In cooking it would be called a recipe, in music a score, and so on. In computing the program performs exactly the same function. Because of this Searle claims quite assertively that the writing of computer programs will never be able to capture the subtlety of human thought.

Given what we have already said about algorithms, this has a superficial plausibility. Computer programs are algorithms. An algorithm is sometimes described as "mindless". Human thought, on the other hand, contains things like judgement, emotion, and understanding. At first glance it certainly does not seem as if this can be reduced to the blind following of an algorithm.

To illustrate this Searle proposes a thought experiment. Imagine a closed room with just a slot to post sheets of paper in and out. Inside the room is Professor Searle together with a vast set of instructions written in English. Somebody passes a piece of paper covered in squiggles into the room. Searle consults the book of instructions. It says that if this set of squiggles is put into the room then a piece of paper with a different, but precisely described set of squiggles must be passed out of the room. This he does. Moments later another set of squiggles comes in; he looks this set up in the book of instructions and passes out another set and so on.

Completely unknown to Searle, the sets of squiggles coming into the room are Chinese characters representing questions written in Chinese. The sets of squiggles being passed out of the room are appropriate answers to those questions. People outside the room are saying that because the room can answer any question posed in Chinese, then it must understand Chinese.

No it doesn't, says Searle. He, John Searle, does not understand any Chinese at all. He is simply following the instructions. This is, says Searle, exactly what occurs in a computer. The computer mindlessly follows a set of instructions – a program: there is no understanding, nor could there ever be with this approach to AI.

Obviously the setting and props of this thought experiment are designed to make us reflect back to the Turing test described in chapter 1. Turing thought that when (and for him it was only a matter of time) a computer could answer an open-ended set of questions in a way indistinguishable from a human then people would say that it could think. He did not see the need to mention "understanding" or to set any technical restrictions on how this performance was to be achieved. Searle's Chinese Room clearly passes the Turing test for understanding Chinese but it does so, Searle claims, without actually understanding any Chinese at all.

Let's look carefully at exactly what Searle is saying here. We are meant to take the Chinese Room as a computer system of which it might be claimed that it can understand Chinese. Searle himself represents perhaps the CPU (Central Processing Unit). The book of instructions represents the program. In this case it is just a vast collection of responses to whatever is input. Real computer programs tend to be far more subtle and flexible than this, but that does not substantially affect the argument. Because of the mindless rule-following character of algorithms, it is claimed that computers can achieve *apparently* intelligent performance without any *real* intelligence.

Many people in AI have responded to this thought experiment over the years, but Searle remains obdurate, dismissing them as "all wrong". Some of those responding think that the "Chinese Room" needs to be put into a real robot in the real world. We will look at this approach in more detail later in this chapter. The most popular response and the one I wish to consider in more detail is called by Searle "The Systems Response". The main claim of this response is that although the Searle in the room does not understand Chinese, the system as a whole does understand Chinese. The systems reply is obviously wrong also, says Searle, because just as there is no understanding in the Searle in the room, there's no understanding anywhere else in the system either. The book of instructions is just a book and the room is just a room. Nothing in the system understands Chinese any more than Searle does so it's pointless to look for understanding in the system.

In this response we can begin to see the sleight of hand that is being perpetrated by this thought experiment. It is not surprising that we cannot find understanding in any of the *parts* of an understanding system. In 1714 the philosopher Gottfried Leibniz famously said that if we imagine a thinking machine as large as a windmill and we enter this imaginary thinking windmill and walk around we will "only find parts that push one another, and we will never find anything to explain a perception".[1] So Searle's observation is hardly new. If we look at the parts of any system that understands Chinese we will not find a particular part that performs that understanding.

The Chinese Room thought experiment deflects our attention from this simple truth by putting a Berkeley philosophy professor into the room. Surely, if we are to look for understanding anywhere in the system it is to the professor that we should look. However, even the Berkeley professor famously does not understand Chinese. Then again, should this surprise us? If we take apart the brain and the body of a native Chinese speaker we will not find one particular part that does the understanding.

More interestingly, let's ask Turing's 1950 question about the Chinese Room.[2] What would we say about a room that can answer absolutely any question? To many people it will seem perverse to say that it doesn't understand Chinese. You can ask any question in Chinese and get a plausible answer in Chinese – does it matter that we can't find the part of the room that does the understanding? Should we care? We don't tear humans or animals apart looking for understanding: we judge them by their performance. It would be rather like saying that an electronic calculator doesn't calculate. If you want to say that then you probably want to reserve the word "calculate" for what humans do. That's unreasonably human-centred. The calculator may not calculate the way you do, but it seems simplest and clearest to say that it calculates. Turing was concerned with people's attitudes in his 1950 paper. It is far from clear that most people would share Searle's reservations. It seems more likely that Turing's prediction would be correct and that the usage of the word "understanding" would subtly change.

## what machines can't do

Another type of opposition to AI has come mainly from mathematicians – not all or even most mathematicians but one or two famous examples. John Lucas and Roger Penrose are names you can pursue in the further reading section. One of the main reasons why these mathematicians find AI implausible is that it has been known for some time – since before there were any modern computers in fact – that there are some things that no computer will ever be able to do. In formal terms, it has been shown that there are some mathematical and logical truths which cannot be arrived at by following an algorithm, or step-by-step procedure.[3]

Now what these writers conclude from this purely formal truth is that human thought cannot be algorithmic. One of the reasons why this view may be attractive to some mathematicians is that they do not *feel* that they are following an algorithm when they make creative leaps in their own mathematical work.

Most people working in AI feel that these views are misguided and they believe this for a collection of reasons. Possibly the most important is that the nature of these formal mathematical and logical truths is so general that it seems most unreasonable to say that human thought is somehow exempt. There are many logical puzzles which stump humans and which are closely related to those which stump computers. One problem which was also known in ancient Greece is that of getting any meaning out of the statement: "This statement is false". If it is true it must be false and if false, true therefore false and so on. For practical purposes, the existence of this and similar paradoxes is not a great problem for humans. Knowing that there are similar and related problems for computers has not stopped us applying computers to an enormous variety of tasks over the last fifty years.

Another reason why people in AI are not usually detained by these objections is that they often seem to make a mistake about the role of the computer in AI. As we saw in the first chapter, the fact that something can be simulated on a computer does not mean that it is itself computational. Simulations of weather patterns, which are the key to modern weather forecasting, must, at the bottom level, use algorithms (or step-by-step procedure) since they run on digital computers. However, nobody is claiming that this means that weather is algorithmic in some simple sense. The simulation is possible because of a scientific claim that there are general laws underlying weather patterns and it is these general laws which can be simulated by algorithmic methods. If we are prepared to entertain similar general scientific claims about human thought – which might be as basic as "human and animal thought is not magical in nature" – then it should be possible to simulate much, if not all, of it on a machine whose underlying working follows algorithms.

## some promises

### putting the robot in the real world

In the 1940s and 1950s the field of cybernetics placed emphasis on building real physical machines such as robots. For most of the subsequent decades this approach was overshadowed by approaches to AI which relied mainly on writing programs, many of which solved fairly abstract problems, like playing chess for example. In cases where researchers were interested in real-world problems (in fairness, this was probably the majority of cases) then a computer simulation was considered perfectly adequate. The researchers of the 40s and 50s did not have the easy access to computing power that became usual for later researchers. Later generations became adept at writing and using effective computer simulation as a research tool with the result that they were much more prepared to tackle problems in simulation, rather than in the real world.

The advantages of dealing with problems in simulation are many and obvious. To alter the design of a real robot may take days or weeks; to alter the design of a simulated robot takes seconds or minutes. By the same token, if the goal of AI researchers is understanding of the principles behind intelligence then the practical problems of assembling and wiring a physical robot are, at best, a distraction and waste of time.

However, during the last decade of the twentieth century a significant minority of researchers began to challenge these assumptions. There was a return to the long-neglected discipline of robot-building. This movement is often called "situated robotics". Given that computer-simulation is so effective, the motivations for this movement need a brief explanation. Why go to all the trouble of dealing with the many practical engineering problems of actually building the robot?

There are, in fact, several reasons why it might be worth the trouble of actually building something. Firstly, simulation can never be guaranteed to be completely accurate. Important features of the problem may be omitted. The further one ventures from what is well-known, the more likely this is to be the case. In the case of much AI research, so little is known that the simulation often becomes a caricature – a gross over-simplification of the complexity of the real world. Secondly, putting the robot in the real world puts a discipline on what sort of control structure is suitable. That is to say that designing the sort of program which is needed to control the robot is no longer such an open-ended problem. The real environment of the robot gives much more restriction than an entirely theoretical one. Proponents of this view stress the danger of deciding how the robot is to behave *before* placing it in a real environment. If we build the robot to operate in an environment then the environment tells us much about its design.

Also, this approach means one can avoid (at least at first) the problem of working out in advance how much the robot needs to know – the problems discussed in the second section of this chapter. Placing the robot firmly back in the real world means that one can postpone problems about what it might or might not need to know. This is captured in the slogan: "the world is its own best representation". We could, for example, tweak the performance of the robot in your living room until we considered its cleaning performance to be adequate. If bumping into the table turns out to be a problem, then we can give the robot a sensor – maybe something as simple as a switch – at exactly the right position to detect the table and turn away. In this case, even asking what the robot knows in order to get the job done might be superfluous. The proponents of this view might well argue that the problem of what the robot needs to know is a misguided philosophical problem which can be avoided by their approach. Once we are satisfied that we have modified the robot's behaviour so that it can get the job done, we don't need to ask difficult questions about how it gets the job done. We don't even have to talk about it *knowing* anything.

## holism

Placing the robot back in the real world also enables a change in emphasis which might be called "holism". Previous approaches to AI had tended to assume that the tasks an intelligent entity might perform could be separated – what is known as "functional decomposition". The situated and embodied approach puts all these tasks back together again. It stresses such things as a close coupling between sensory input and motor activity. That is to say that there should be a minimum of computational processing in such a sensory-motor loop. This is something often found in nature.

For much of the history of AI it was assumed – sometimes explicitly, sometimes implicitly – that intelligent behaviour could be tackled piecemeal. Some researchers worked on computer vision, others on planning, others on language and so on. While this still continues in some areas of AI, it has come under attack during the last ten years by researchers who believe that the problems of intelligent behaviour cannot be broken up in this way.

One of the leading exponents of this "holistic approach" is Rodney Brooks, director of the AI lab at the Massachusetts Institute of Technology (MIT) – perhaps the most influential position within AI. In an important paper published in 1991,[4] Brooks uses what we might call the parable of the Boeing 747 to illustrate this point. Imagine, he says, a group of researchers working on the problem of artificial flight in the 1890s being taken a hundred years into the future in a time machine. There they are given a ride on a Boeing 747. They go back to their own time full of enthusiasm, for, having seen the future, they can be sure that artificial flight is possible. However, everything else they learn from their experience is a big distraction.

The infant science of aerodynamics is abandoned, for example. During the flight they quiz their fellow passengers on this field of science and, finding total ignorance, conclude that it must have led nowhere. A group of researchers start to copy the seats. At first they use hollow steel tubing (the aluminium from which airline seats are actually made could not be refined in the 1890s). When someone observes that weight cannot be a problem if something as obviously massive as the 747 is capable of flight, they see the folly of this approach and use solid steel bar instead.

One group has seen an engine with cover removed and is inspired to try to copy this. However, the modern high bypass turbofan engines which power a 747 are way beyond the technology of the 1890s in just about every respect. The materials are not available. The engineering precision is not available. The knowledge of gas flow required to design such an engine is not available, and so on.

I call this a parable because Brooks offers no explanation of how his "story" is to be read. However, the obvious reading is that the 1890s artificial flight researchers represent traditional AI researchers and the 747 is the human mind. It is inspirational to be given this glimpse of what intelligent performance can do, but almost all of it is so far beyond existing technology that it can only mislead. Those who try to reproduce the modern jet engine are the neural net researchers. Inspired by brains, they try to copy them but they are as far beyond existing technology as turbofans would be to 1890 aviation pioneers. Those who try to reproduce the seats are those who believe that by reproducing small parts of human ability – say, the ability to put a sequence of actions into a plan – they are attacking the general problem of intelligence.

Through this parable, Brooks is, of course, being highly critical of functional decomposition – the idea that intelligence can be tackled in separate distinct research programs. What he is arguing for is a holistic approach – building simple, if not downright crude, embodied robots in order to research all the problems of general-purpose intelligence.

The 1890s artificial flight researchers eventually succeeded by pursuing the science of aerodynamics and by using sound aerodynamic data to build very crude aircraft from the materials then available. The Wright Brothers flew the first powered aircraft in 1903 as a result of diligently pursuing these methods. These early beginnings scaled up remarkably quickly. In fact, only sixty-six years later – within a single lifetime – Neil Armstrong stepped on to the surface of the Moon. More importantly for present purposes, in that same year, 1969, the first Boeing 747 – the aircraft that has come to characterize modern travel and forms the basis of Brooks' parable – first took to the air.

It is too soon to say if the holistic and embodied approach of situated robotics is about to pull off something equally spectacular in the field of AI. However, this is perhaps a promisingly humble way of looking at some AI problems. The possibility that something as spectacular as the development of aviation could be achieved in the next sixty-six years cannot be ruled out.

## artificial life

Putting together the idea of building "situated robots" and a holistic approach to AI led, in the early 1990s, to the foundation of a new approach to AI (some might want to say a new discipline) known as Artificial Life. One of the main reasons that the proponents of this new approach to AI wanted a new name was their outspoken opposition to previous approaches to AI. This sort of factionalism is nothing new, as we have seen. We might even allow ourselves a brief chuckle in that the name "Artificial Life" continues squarely in the tradition of previous approaches, namely that of sounding far grander and more general than is actually justified. That said, Artificial Life seems a very promising way of looking at the problems of AI. The notion of situated embodiment is not only a good way of building robots. It also helps to deal with some of the more general problems that have been holding back AI.

Unfortunately, after 10–15 years of development it has become clear that artificial life fails to "scale up" in a very similar way to previous approaches to AI. Situated robotics has produced some wonderful machinery but situated robots tend to perform well only in a very specific situation. It looks like this is not an easy route to building general-purpose intelligence. Of course, to say that situated robotics is not an easy route to general-purpose intelligence is not to say either that it is not a route at all or that the easy route is to be found elsewhere. It is, like the other approaches we have seen, a part of the jigsaw. It's just to say that it does not look like it is the only part.

I am reminded of a cartoon by Margaret Welbank that appeared on the cover of *AISBQ* a few years ago. *AISBQ* stands for *The Quarterly Journal of the Society for Artificial Intelligence and the Simulation of Behaviour* and it is the trade journal of the British AI research community. The cartoon represented various approaches to AI as human characters. The approach based on search and heuristics, described in chapter 2, was shown as a bearded old man. The approach based on neural nets and inspiration from brains, described in chapter 3, was drawn as a mature net-wielding individual. Both these figures were looking at artificial life – portrayed as a crawling baby. "Just wait till he's had a few failures" they were wryly observing.

Now that artificial life has had a few failures we can perhaps see the position a little more clearly. Situated embodiment and a holistic approach avoids some of the puzzles that other approaches were stumped by. What it doesn't, at present, seem to do is to scale up to enable us to build general-purpose intelligence.

One of the best demonstrations of this failure to generalize is what many would regard as the "flagship project" of this approach to AI. This is the humanoid robot known as Cog built at the MIT AI Laboratory. This is the best-funded AI research centre in the world and Rod Brooks is, as we have seen, an outspoken advocate of this approach to AI. Cog is a humanoid robot in that it has something rather like an upper torso, fully articulated arms, and a mobile head and eyes (see opposite).

The reasons for building the robot like this are as we have already discussed. The team at MIT believed that rich sensory and motor connections with the world were necessary for the development of the right sort of behaviour by the robot. The key idea is that its perception involves not just taking in abstract information but rather it involves a rich set of connections between perception and action – what its builders call the sensorimotor loop. In short, only by making the way the robot interacts with its environment sufficiently similar to the way a human interacts with its environment, could they ensure that it would acquire human-like concepts. Furthermore, in the case of

*Cog – the humanoid robot built in the AI lab at MIT (©MIT Artificial Intelligence Laboratory)*

Cog, it was thought that humans learn many of their crucial skills through interaction with other humans and that Cog would develop through such interactions. Therefore it had to be able to look at a speaker, return their gaze, and even hold hands.

Early successes abounded. Cog learned to reach its hand to a visual target in the way that very young human infants do. Cog can track objects and sounds. However, the "higher level" skills that all this machinery and development is supposed to underpin have not emerged. Rod Brooks gave one description of this failure in saying that, despite all the effort, Cog still cannot tell a cell phone from a glasses case. It is important to remember that Cog is a complex robot which is at the focus of many different research programs and the failure to meet this one particular research goal does not mean that the whole approach is in difficulty.

Opinions vary greatly as to why the object recognition problem persists with this approach. At one extreme are those who feel that Cog simply needs more time and more money, at the other extreme are those who feel that the whole idea was misguided in the first place. Between these extremes some more realistic comments can be made. Firstly, there may be organizational and political difficulties in managing a project as large as this. Since about fifty per cent of all IT projects fail, we should not read too much into any individual failure – it only had an even chance of success in the first place. Even more importantly there was just too much "missing science" in the gap between low-level technical skills like grasping an object and things that we might want to claim were the beginnings of intelligence – things like recognizing particular objects. It is one thing to express a fervent commitment that such things can only be achieved by a holistic, situated approach exemplified by Cog, and quite another to claim to know *how* this can be achieved. Finally this might be a scientifically very illuminating failure. This is a point to which we will return when considering future trends in the final chapter.

This problem of "missing science" is an important one for the fast-maturing field of Artificial Life. There is still an awful lot we do not know about the biological processes that underlie natural intelligence. We do not yet have a complete picture of how a single neuron (brain cell) operates. We do not fully understand what happens at the synapses (junctions) through which neurons communicate. We do not yet understand how the various chemicals which flow through brains affect their performance. Artificial Life researchers have spent much time and effort looking at fairly simple animals such as insects. One of the conclusions that they have reached is that these supposedly simple animals demonstrate remarkably complex behaviour – usually too complex to be reproduced in a robot. This is an example of the way in which one learns awe and wonder at the natural world through trying to understand it. It is also an example of the problem of "missing science".

This means that the new field of Artificial Life is already in crisis and may soon split into two further, rather different, areas. One, very concerned about the missing science, will see the proper goal of researchers as helping biologists find out more about this missing science. Computer simulation of simple animals such as insects will be used primarily as a route to achieving better understanding of the behaviour and biology of such animals. The other group either think or hope that the missing science does not affect their goal of building a robot with general-purpose intelligence. They will continue to build and hope that something interesting will emerge. I do not mean to sound critical by using the expression "build and hope". It is important to remember that sometimes it is possible for practice to precede theory in science. Cog, at least in the object-recognition case, is a prime example of this "build and hope" approach.

There's even a lone British researcher – Steve Grand – attempting to build an embodied robot with mammalian-style intelligence (see picture, p.89). He is doing this totally

independently in a workshop built by converting his garage.
What is more, Grand has no academic, commercial, or military
funding. This approach may sound rather over-ambitious when
put alongside, say, the Cog project with its massive funding.
However, readers should recall that lone British inventors have
pulled off surprising coups in the past. John Logie-Baird, for
example, built the first working television working alone in a
small flat in Hastings on the south coast of England.

Grand is just such a highly individualistic researcher who just
might pull off something spectacular. He should certainly not be
seen as an enthusiastic amateur. Not only is he familiar with the
state of the art at various research centres but he is also well
known within AI. His previous claim to success was the
development of a computer game called Creatures, in 1996. Not
only did this game introduce a new genre of computer games
(now known as "bring-em-ups" rather than "shoot-em-ups")
but it also showed how evolutionary computing techniques
could be used in an entertainment technology.

Most people in AI wish him well, though they might be more
cautious in their estimates of his chances of success. What he is
trying to do is extremely difficult. Many informed commentators
would call it impossible. Steve Grand is an illustration of the
claim, made at the start of the chapter, that intelligence,
creativity, and persistence are the only qualifications required to
participate in AI research.

## further reading

A witty and readable account of why the frame problem
matters can be found in *Cognitive Wheels* by Dan Dennett
(1984).

A non-technical introduction to what philosophers have to say
about some of these questions is Tim Crane's *The Mechanical
Mind* (1995).

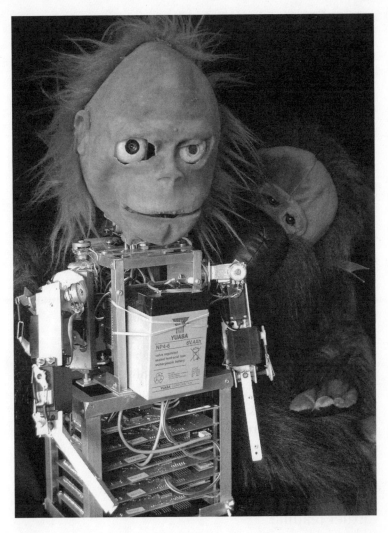

*Steve Grand's robot, Lucy*

Andy Clark explains the 007 principle in *Microcognition* (1989) but his later book *Being There* (1997) would perhaps make a better companion to this chapter.

John Searle describes the Chinese Room thought experiment in a number of places including 'Minds, Brains and Programs' (1980); *Minds, Brains, and Science* (1991); and *The Rediscovery of the Mind* (1994).

John Lucas made his original claim in a 1961 paper which is included in Anderson, 1964 in the bibliography. Roger Penrose sets out a more detailed and wide-ranging attack on AI in *The Emperor's New Mind* (1989).

For an introduction to the enthusiasm surrounding Artificial Life the best book is Steven Levy's *Artificial Life, The Quest for a New Creation* (1993).

The paper in which Rod Brooks uses the 747 parable ("Intelligence Without Representation", 1991) is available online together with many details of the Cog project at: http://www.ai.mit.edu/people/brooks/

Rod Brooks's latest book, *Robot: the Future of Flesh and Machines* (2002) is a general-audience account of his approach and work.

Steve Grand sets out his stall in *Creation: Life and How to Make it* (2000). You can keep up to date with Steve's progress at: http://www.cyberlife-research.com/people/steve/

## notes

1.  "... if we imagine that there is a machine whose structure makes it think, sense and have perceptions, we could conceive it enlarged ... so that we could enter it as one enters a mill. Assuming that ... we will only find parts that push one another, and we will never find anything to explain a perception". (Leibnitz, *Monadology,* Sec. 17).

2. See section on "The Turing test", chapter 1.
3. The most important for practical purposes is the so-called "halting problem". This states that it is impossible to determine in advance whether or not a given computer will halt when running a given program with a given set of inputs, or whether it will "loop forever". It is important to remember that this is something that has been proved mathematically, rather than found out by programmers; indeed it was known before there were any programmers!
4. Brooks, "Intelligence Without Representation", pp. 139–59.

# ai diffuses

## the "yes, ... but question"

When you give a talk to non-specialists on AI, draw to a close, and ask if there are any questions, then the most likely (and probably the first) question to be asked will be what I call the "Yes But Question". Often it will take the form of something like, "What you have said is very interesting... *but* when are we likely to see computers that are *really* intelligent?" This is a difficult question to answer directly. Respect for one's audience usually demands something that politely starts "Let me explain again that...". There is clearly a gap of understanding that needs to be bridged here.

It really does seem that the Yes But Question is one of the main things that the public wants to ask of AI commentators. I am therefore going to try and answer it directly and fully. As we have seen, AI research is progressing steadily and producing many useful products and many exciting ideas. If that is not what you mean by computers that are "*really* intelligent" then what else might you mean? Well, one guess, informed by experience, is that you mean intelligence that is *completely* human-like. The Yes But Question might mean something like, "When will I meet a computer that thinks just like me?" There are quite a number of things to be said in response to this.

As we saw in chapter 1, human-like intelligence has sometimes been a goal for AI. Some interpretations of the so-called Turing test imply that imitation of human intelligence is a useful goal, if not a sort of "gold standard" for determining whether or not a computer is intelligent. But I have also described the inadequacy of definitions of AI which specifically compare machines to humans. At the moment we know so little about human intelligence, other than that it is truly wonderful, that this is unlikely to help in the development of AI.

I suppose a fair (if brusque) answer to this particular reformulation of the Yes But Question would be something like, "Given that we have no useful understanding of human intelligence at present (remember Brooks' 747 in chapter 4), who knows? And it's not clear that our aim should be to build a computer that thinks exactly like you or me anyway."

AI often deals with rather different forms of intelligence and there is, of course, no shortage of human intelligence. Of course, saying all this in answer to the Yes But Question can lead some people to believe that what is being talked about is not intelligence at all. "Either it's like me, or it's just a machine following instructions" they often seem to be saying. We have already seen that the common sense view of a machine is misleading. The machines that I am talking about in this book most certainly do not "just follow instructions". Not only is it possible to talk of them having goals, it is also realistic to talk of them selecting between different goals and responding differently in different situations. They contain and use knowledge and are capable of learning. Many people within AI – and outside – feel that they do these things in a fairly limited way but that makes no difference at all to my claim here. What they most certainly are not doing is "just following instructions".

One of the best illustrations of this is a passage in a book by Sherry Turkle (Turkle, 1984). Sherry Turkle is interesting for the present discussion because she is a practising psychotherapist as well as having an interest in AI. At the time of the story she relates she was also involved with one of the leading AI researchers at MIT. She describes how one morning she sees a patient who is very depressed because he feels that he is "like a machine". Yet later that same day she goes to a party for students and researchers in the AI lab at MIT. There she meets a young woman who, growing inpatient with a discussion about whether machines could ever think, says: "What's the problem? – I'm a machine and I think". This student talked about being a machine as a positive and liberating way of seeing oneself.

The reason for the two very different reactions, she says, is because there are two very different meanings of the word "machine" being used here. Her depressed patient uses the words "like a machine" to illustrate a life over which he feels he has no control. It is a vision of a nineteenth-century machine, its parts moving in simple predestined arcs. He feels he has no choices; the gears just turn. The AI researchers and their students, on the other hand, mean something very different when they talk about machines. They mean the sort of machines we have been considering in this book. These machines have multiple goals and can choose between them. They may choose different methods to achieve those goals and can negotiate with the real world – including other people – to achieve these goals. This is liberating because it shows one way in which our choices could make a difference, even in a deterministic scientific universe. AI opens up the possibility of machines that do much more than just follow instructions.

So one meaning of the Yes But Question may be nothing more than a refusal to accept the possibility that there could be intelligent entities that are different from ourselves – a refusal based on a nineteenth-century picture of a machine. Such a refusal may be mistaken and misleading. Perceptive readers may now become impatient that I have talked of intelligence without defining it throughout this imagined debate. There is a discussion of the definition of intelligence in a later section.

In an important sense all the preceding chapters are a response to this interpretation of the Yes But Question. We have seen that AI has produced and will continue to produce impressive working technology. We have entered the age of smart machines. We have seen that critics who said that things would never be done have been proved wrong. Fascinating insights into how human and animal brains work have emerged.

Yet people still want more. It is not at all clear how science could provide more. Many people will mistrust the intrusion of science into this area at all. One motivation for mistrust is the lingering belief that human intelligence is somehow special – something that just never could be understood by science. For others it is a sort of battle over boundaries. Poets and novelists have been the main authorities on the subtleties of human thought throughout history and they may well resent the idea of scientists moving in to clinically measure, to simulate, and to build machines on what they perceive as their territory. However, there are a couple of more sober and serious reasons for mistrusting science in this particular case.

These more serious reasons for mistrusting the scientific approach can be divided into two main areas. The first area is the so-called "problem of consciousness" – actually it's not agreed that it is a problem and hardly anybody thinks it's just one problem. If the Yes But Question includes the "consciousness" word then it's very hard to give a quick answer. Consciousness has become a very large field of study these days and will be considered in a later section. The second is the essential subjectivity of items such as thought, beliefs, and so on. These areas are a reason for looking at the Yes But Question in more detail here. They may, perhaps, be the reason why some people feel the need to ask the question in the first place. We'll have to digress a little before looking at these two areas later in the chapter. The digression will be worthwhile as it will take us through a whole new area of science.

## cognitive science

### folk psychology

Psychology is in a curious position as a science. Modern psychologists tend to find themselves in a similar position to mediaeval geographers. If their scientific research confirms peoples' prejudices – that the world is flat – then they will be attacked for spending time and money finding out the obvious. If their research discovers something which goes against general common sense – that the world is round – then they will be dismissed as out of touch with reality.

There are a number of reasons for this but one of the most important is that in everyday life we all use something that superficially looks rather similar to psychology to allow us to understand and predict what other people will do. Philosophers call this "folk psychology". There's a good deal of debate about what exactly folk psychology is and about how much it matters. For the moment we can say that we ordinarily talk as if people usually do things because they have desires and they believe that by doing these things they might achieve those desires. Why are you reading this? Well, in folk-psychological terms, you have a *desire* to learn something about AI or even cognitive science and you *believe* that by reading this you will find out. (There's no guarantee that any of your beliefs will turn out to be correct of course.) If your next action is to go to the kitchen and get the coffee pot out of the cupboard then a good explanation might be that you want a cup of coffee and you believe these actions will help you get it.

Folk psychology forms the basis of how we normally get on with other people. This world of thought and belief can become much more complex than the minimal picture I have just given. It forms the basis of art and literature. Indeed we could say that drama is just the conflict of desires and beliefs either between people or within the same person. Folk psychology is also the basis of our various systems of rules and punishments. The problem is that it does not seem to be terribly scientific. If you take brains apart, or examine live brains in a scanner, you never find any beliefs. We can measure what people actually do but the only way to find out about their beliefs is to ask them. Even "common sense" tells us that this is not reliable. We find many differences between what they say their beliefs are and what they actually do. To say that people *obviously have* beliefs is just like saying that the Earth is obviously flat. Proper scientific measurement can show that it is not actually flat. For this reason scientific psychology must treat folk psychology with extreme caution.

Philosophers all agree about the importance of folk psychology but there the agreement definitely ends. Some say it is a full-blown psychological theory and some say that the job of scientists is to ground it in other sciences – for example to show how the electrical and chemical processes in the brain could allow the brain to form beliefs and desires. Some say that it only looks like a scientific theory and really it is little more than the ability of humans to empathize and put themselves in someone else's position. Still others say that when we have a truly scientific psychology then all this talk about beliefs, desires, and even minds will disappear in the same way as did the flat earth view in geography. There are many difficult questions involved with this.

Our main concern here, however, is the way in which AI fits into the story. In the first half of the twentieth century the main scientific challenge to folk psychology and common sense came from an approach to psychology known as "behaviourism". This approach dismissed common sense and folk psychology as hopelessly unscientific. Instead it claimed that the only *scientific* way to explain human behaviour was to analyze it in terms of a combination of input and outputs. People were subjected to a "stimulus" and then made a "response". Scientific psychology was nothing more than measuring and recording these stimulus/response pairs. Any talk about what people might believe or want or think formed no part of a scientific psychology, said the behaviourists. At best, they said, the human brain is a "black box". Things go in and things come out (i.e. stimuli and responses) which scientists can measure. Talk about what goes on in between stimulus and response was not the concern of scientific psychology, they claimed.

Cognitive science was born partly out of opposition to behaviourism. AI was crucial in allowing that birth. The use by engineers and AI scientists of much richer descriptions about what goes on inside machines showed that even computers were not "black boxes" in the behaviourist sense. For example, they often talked of a program or robot having a "goal". This would be something that it was trying to achieve and which showed in its actions but also something that could be pointed to in the inner workings of the program or robot. Suppose you sent your domestic robot to the kitchen to get you the coffee and its internal registers therefore contained the "make coffee goal". If we then ask why it went to the cupboard and took out the coffee pot then the answer is not a million miles away from the folk psychological explanation we would give for a human.

This opened a door, and a very important door. If people in another branch of science use certain terms in a way very similar to the way in which we use them about humans then the claim that these terms are "totally unscientific" is undermined. At the very least we must talk of computers as information-processing devices. If we allow this small step with machines then we don't have to treat humans as black boxes any more. It becomes scientifically acceptable to talk of them as information-processing devices.

After this small step in freeing psychology from over-zealous behaviourism, the relationship between AI and cognitive science became much more complicated. Many people got a bit carried away at this point so some cautions are in order. Allowing that it is *scientifically acceptable to describe* humans as information-processing devices does not mean that they actually *are* information-processing devices. This remains to be shown. Even if it were shown it would not mean that this is *all* they are. Still less does it mean that the beliefs and desires of folk psychology are directly equivalent to bits of a computer program. This inspiration from AI is known as the "computational metaphor", and it is important to remember that it is just a metaphor.

## what cognitive science is

It's very hard to say what cognitive science is. The disagreements about the scope and methods of this new science are, however, a very healthy situation. Thomas Kuhn, in one of the most famous books about science,[1] distinguished what happens in a scientific revolution from what happens in so-called "normal science" – when everyone is just plodding along. When a scientific revolution occurs – what Kuhn calls a change of paradigm – scientists do not just find different ways of researching their area. What also happens is that the meaning of crucial terms changes too. So, for example, when physicists changed from the paradigm of Newton to the paradigm of Einstein they did not simply decide that the universe was different from the way they previously thought it to be. Some of the most important words in physics like "space", "time", and "mass" changed their meanings. In the case of cognitive science the revolution is occurring right now and many of the most important terms are still in the process of being defined. That is one important reason why I have put off giving any definition of intelligence to this chapter. If I had given a definition at the beginning that would commit me to just one view at the outset, the result would have been a very different book. It would certainly not have resulted in a tour through some very different ways of looking at the area.

I hope now that readers will understand why there is much disagreement about exactly what cognitive science is – and even more about what exactly it studies and what is the correct way to study it. This disagreement is not, however, symptomatic of a subject area in a mess. It is symptomatic of a science in the excitement of creation of new ideas and new ways of looking at the world. It also means that cognitive science is, just like AI, a cross-disciplinary activity. It involves (at the very least) psychologists, neurobiologists, linguists, computer scientists, and philosophers.

Cognitive science also has the best claim to be the science of mind in humans, animals, machines, and extra-terrestrial aliens (if there are any) which I said was needed in chapter 1. In this respect cognitive science is in a position analogous to that of aerodynamics in the case of artificial flight. As an infant science, it can't really tell us that much about the aerodynamics of intelligence yet but the possibilities are great. As an historical aside the science of aerodynamics and the engineering enterprise of building aircraft progressed side-by-side. The crucial scientific paper which allowed the development of modern aerodynamic theory was published by a German mathematician – Ludwig Prantl – in 1904. This was just one year after the Wright Brothers' first successful flight.

There are many potential practical benefits which might stem from the development of a general scientific account of intelligent behaviour – in addition to satisfying scientific curiosity about just how human intelligence works, of course. As we come to understand the science of intelligent behaviour better we may well be able to build better interfaces to computers, or any other machines for that matter. This will be of great benefit to humanity, and to me personally. At present I am using three devices – QWERTY keyboard, mouse, and monitor which are *known* to cause injury in humans even when used correctly.

If we understand more about the science of intelligent behaviour we may well be able to improve the technology we use to help people in decision making. An important area of AI research is known as "decision support systems". Since we often all suffer from bad decisions by managers and other leaders and often all benefit from good decisions, we should very much welcome anything that AI can do to help here. The same is true for technology used to support teaching and learning. At present it often fails to live up to expectations. The reasons for this are, as we have seen, that teaching and learning turned out to be far more complex than had been imagined before anyone tried to build AI technology to do it. Cognitive science promises to throw light on the process of teaching and learning which may well result in more effective technology.

Cognitive science has also had a tremendous influence on human psychology. Although it is usually only dated from the 1970s it has, in the English-speaking world at least, come to be the dominant approach to scientific psychology. Mainly because of this it has become the most important way in which AI has affected science and society by exporting ideas, rather than by producing working technology. AI has produced technology but the export of ideas to other disciplines is, at least, probably more important.

From a personal perspective it seems that we are very much at the start of things. The way I characterized cognitive science as an infant science – or still in the throes of Kuhnian revolution – affects all of AI. It's an exciting time to be involved.

## what about the turing test?

In the original paper, Alan Turing clearly stated that by about the year 2000 computers should be able to play the imitation game so well that the average interrogator would have only a thirty per cent chance of determining which is the human and which is the computer. If you follow the URL given in the further reading for chapter 1 you'll see that the entries for the most recent Loebner Prize (an annual Turing test-like competition) are superficially impressive but nowhere near passing a full-blown Turing test.

The Turing test is not a central focus of current AI research by any means and there are good reasons why it should not be.[2] However there are reasons why AI research may keep returning to human intelligence. The most important of these is that for some people in the field of AI and most people in cognitive science the most interesting problem simply is the problem of understanding human intelligence.

If AI tells us anything about this problem it is that our understanding of our own intelligence is extremely limited. Just because we use it everyday doesn't mean we have much insight into how it works.

## what about consciousness?

When we looked at the Yes But Question two problems were postponed. They were the problem of subjectivity and the problem of consciousness. For the sake of completeness this section attempts to give a brief picture of the enormous and deep debates on these two questions. Since the section is, like the debates themselves, highly inconclusive, readers with no particular interest in these questions may safely skip it.

There are three ways we use the word "conscious" in normal everyday talk. The first is roughly to mean awake as opposed to asleep – what an anaesthetist means by conscious. The second is roughly to mean self-aware as in, "At this point I became conscious of just how anxious I was about landing in the weather conditions prevailing at San Francisco". The third is altogether more difficult and is often heard in relation to the Yes But Question with which we started this chapter, as in, "You may be able to build all those abilities in to a robot but it still won't be conscious."

The first two meanings of consciousness don't seem to pose any great difficulty for AI – at least in principle. Something roughly along the lines of switching the robot on and off will satisfy the anaesthetist's definition. For self-awareness we can say that the computer I am using now contains programs which look at other programs and can show what proportion of time it is spending on various processes and so on. In principle then, building self-awareness has already been achieved. This is "in principle" because, as we have seen, there are many practical problems that AI researchers have to solve to deal with the first two meanings of conscious. However it does seem at least possible to deal with them. Indeed, the example given in the second is chosen particularly because the decision not to land at San Francisco in the bad weather and to return instead to Merced is just the sort of decision that modern decision-support systems are aimed at. In flying (and many other areas) it's often important to take a decision quickly but also to take account of a large amount of diverse relevant information. AI is making steady progress at automating such decisions.

The third meaning of conscious is more difficult even to state in simple terms. In 1974 the philosopher Thomas Nagel published a landmark paper in which he argued that there was something about our subjective experience which science could never capture.[3] In vague terms this something is the experience of what it is like to be you, just you, now. The reason, according to Nagel, that science will never tell us anything about this is that science is in the business of finding general laws. What it is like to be you, now, is exactly the opposite of general – it is specific to you at this moment.

There is much debate on this area at present, mainly by philosophers, but neuroscientists, robot builders, and mathematicians like to get involved too. Some people believe that there really isn't any such thing as what it is like to be you at this moment. It's a "useful fiction" that enables you to look after yourself and plan your activities in the world. Others believe that there might actually be something that it is like to be you but that it doesn't actually make any difference to anything. A robot, or an intelligent extra-terrestrial alien, could have all your mental abilities without having consciousness in the third sense. Still others believe that it is real and that it does make a difference. Any reader interested in this area should pursue it through the further reading section as this is only an overall map of a very large and very active debate.

The consequences of this big debate for AI and cognitive science are equally hard to pin down. For each of the positions on whether or not it is real and whether or not it matters, there is a range of views about what the consequences are for AI. Some philosophers believe that because it is real and because it matters, AI is a complete waste of time. Most people in AI would respond that whether or not it is real and whether or not it matters are questions that should be determined by experiment, not by philosophers pronouncing from outside. Some people believe that only things that have the right sort of evolutionary history can achieve consciousness in the way that we humans have done. Many of them are therefore very interested in the simulation of evolution as a way of designing robots, which was discussed in chapter 3.

On the other hand, if consciousness either doesn't matter or is a "useful fiction" then (at least in principle) we can build intelligent machines without being forever detained by the debate over consciousness. However, even here there is not agreement. Some people in AI say that consciousness is an "emergent property". That is roughly to say that although there may be no way in which we could design it into a robot, if a robot is complex enough, or has the right sort of interaction with its environment, or both, then consciousness will simply emerge. Believing this is very exciting as it is the ideal philosophical justification for the "build and hope" approach mentioned at the end of chapter 4. "Build and hope" may sound a slightly critical slogan but this approach is really quite justified. One way of bringing this wide-ranging philosophical debate down to earth would be to conduct some relevant experiments, and that is just what these robot builders see themselves as doing. When Orville Wright made that first little flight at Kitty Hawk in 1903 all the eminent scientists who had said (and were still saying) that it was impossible were shown to be wrong in one decisive experiment. Of course, determining whether or not an aircraft flies is rather more clear-cut than determining whether or not a machine is conscious.

Adding the question of subjectivity to this third problem makes it rather more difficult. If Nagel's account is more or less right then this sort of consciousness can only be experienced from a subjective viewpoint. In other words, only the robot itself can really know whether or not it is really conscious in this sense.

I can make one final comment on this possibility. A few years ago my duties at a British university included shutting down the computers in the AI lab. There was one machine that, whenever I did this, shouted loudly, "No! Don't turn me off, I've become conscious" – I never had a moment's hesitation in turning off this machine, sure that what I was hearing was a consequence of undergraduate humour rather than progress in AI. If any AI researcher made similar claims about subjective consciousness my reaction would be rather similar.

## what's intelligence then?

I promised in several places that a fuller examination of the definition of "intelligence" would have to be deferred to later in the book. I hope that most readers did not cheat and read this first because what I say here really does depend on much of what has been said before. Many important lessons about intelligence have been at least touched upon through the course of the book.

Throughout the book I have used a metaphorical comparison with the history of aviation as a way of understanding what is going on in AI and cognitive science. A general scientific account is now being worked out that will one day explain human, animal, and machine intelligent behaviour (and that of extra-terrestrial aliens if there are any). This will be a general branch of science which will do what aerodynamics did for flight. We now know that birds and insects and frisbees all must conform to the general scientific laws of aerodynamics.

Of course, intelligent behaviour is a lot more complex than flight. Producing this "aerodynamics of intelligence" is not going to be easy. However, although the work is at an early stage, it is very much underway. Because it is very easy to get carried away with the excitement of a vast new area of science much of what I have said and will say is very deliberately over-cautious.

If you have heard of the distinction made between "Weak AI" and "Strong AI" made by John Searle (he of the "Chinese Room" in chapter 4) then, in as much as the distinction makes any sense to me, I have placed this book consistently in the "Weak AI" camp. AI has yielded fascinating insights about human thought and will continue to do so. However, AI and its continuing progress do not entail producing anything that "thinks just like you". Furthermore, I happen to believe that even if (and I have said why I consider it highly unlikely) some artificial device one day reliably passes the Turing test, it would probably still be nothing like you in its thinking.

So why am I using the word "thinking" about this possible future device, you may ask. Well, it's just too arrogant – the technical word is anthropocentric – to reserve such words only for humans and the peculiar way humans do things. Both electronic calculators and humans calculate. They use very different types of material and they perform the task, as far as we can tell, in very different ways, but they are still performing the same task. The same applies to intelligence. Most readers will think of their own intelligence whenever I have used the word, despite my urging them not to do so. Once again, being anthropocentric about intelligence is really very unhelpful. Probably the least helpful way of thinking about intelligence is to think of IQ (the so-called intelligence quotient).

The reason IQ is particularly unhelpful is that it suggests that intelligence is a single quantity – that it is unidimensional. Describing intelligence through a single number value is very misleading. IQ tests also tend to measure abstract problem-solving ability. As we have seen, robot builders tend to be much more interested in the ability to avoid bumping into the furniture than in solving abstract problems. These two very different manifestations are just two small parts of the jigsaw of intelligence. One of the many lessons that can be drawn from AI is that we cannot describe such a complex multi-dimensional phenomenon as intelligence in anything approaching a single number. When we use the word "intelligent" of behaviour we are not talking of a single property.

Abstract problem solving is only once piece of this jigsaw. It is important not to forget the environment and the lessons from situated robotics. A quite different account of intelligent performance emerges if we are more interested in avoiding the furniture than in playing chess. Putting these two accounts into one overall framework is certainly a challenge but it is a challenge that needs to be solved.

Please remember Brooks' parable of the 747. Whenever we talk of intelligence there is a massive temptation to think in terms of human intelligence. One thing that AI has conclusively demonstrated over the years is that human intelligence is truly wonderful. All around us (and within us) we constantly see evidence of its power, its versatility, its creativity, and its ability to navigate around the living room without bumping into the furniture. This is so far ahead of the present state of science and technology that it can do little but amaze and dazzle serious scientists. What we can currently achieve with computers and robots is less in comparison to human intelligent performance than a string and glue kite is to a Boeing 747.

There's an even more interesting reason for not looking too hard at humans. Some accounts of the evolution of human intelligence suggest that much of it has *not* evolved to meet purely functional needs.[4] The reason for this is that human intelligence may be rather like the enormous and spectacular tail of a peacock. The peacock's tail does not help it to fly, fight or feed. Indeed in aerodynamic terms it is a serious hindrance. What it does for the peacock is to help it get a mate. Peahens have evolved to select mates on the basis of the size of their tails. It seems quite possible that much of human behaviour – in particular music, poetry, and so on – was the product of this sort of mate selection. If this is the case then it is not just the power and flexibility of human intelligence that distracts us when building artificial intelligence. Much of what we see when we look at human intelligence is like the peacock's tail – it did not evolve to help us fight, feed, and survive. It is mainly a display for attracting mates. In this case the only reason to research it as scientists is curiosity about ourselves. There's even less reason to imitate it as technologists unless, perhaps, we are interested in producing robots for the sex industry.

Artificial intelligence is genuinely artificial. The small crude attempts at it that have so far been made suggest that it looks very different from natural intelligence. However, it is still intelligence. How closely does a fruit fly resemble a Boeing 747?

## further reading

If you are interested in understanding the differences between human and artificial forms of intelligence then Geoffrey Miller's, *The Mating Mind*, is a very good starting point.

Consciousness has, over the last decade, become a large multi-disciplinary field of study. If you want to look into it further, I'd suggest *Consciousness Explained* (Dennett, 1993) as one way into the subject.

A book which takes a different approach drawing more on human emotion and less on the computational metaphor is *The Search for Mind, A New Foundation for Cognitive Science* by Sean ONuallain (2002).

A good general overview of Cognitive Science and its relationship to AI is *Minds, Brains, and Computers* by Robert Harnish (2002).

## notes

1. Kuhn, *The Structure of Scientific Revolutions*.
2. See Whitby, *Reflections on Artificial Intelligence*, for more details on this.
3. Nagel, "What is it like to be a bat?"
4. This interesting theory has been set out by Geoffrey Miller in *The Mating Mind* (2000).

# present and future trends

This chapter considers first the impact of AI on society as a whole. While there is undoubtedly a need for informed public debate about AI, in general it is a remarkably benign technology. Secondly, it looks at one, perhaps slightly surprising, place where we see a coming together of humans and AI technology, namely the use of AI in art. To round off we can look to the future, at least the immediate future of AI itself.

## the social effects of ai

Almost all technologies change the people and societies that adopt them – often out of all recognition. Most of the readers of this book would be like fish out of water if transported by time machine to a pre-industrial society. Very few of us could hunt and grow sufficient food, let alone build adequate shelter.

Agriculture, roads, telephones, trains, and so on mean that we lead very different lives from our forebears. In an important sense, we are all products of this technology, rather than simply beneficiaries of it. That is to say that we often define ourselves and our roles in terms of technology. It is not just when we describe our activities in technological terms that this is true –

when we say we are driving, sailing, or painting. Much of what we now spend our time doing would simply not make sense to inhabitants of primitive societies. We tend to watch a lot of television, keep in touch by phone and email, and sell a lot of insurance.

AI is likely to have just as big an impact *as a technology* as have these previous technologies. One important thing that is different about AI is that it is also having an impact on the way we think about ourselves. The way AI has exported ideas and its distinctively different way of looking at the world was the theme of the last chapter. However this distinctive mix of both ideas and technology also characterizes the effects of AI on society as a whole.

## the social effects of ai technology

There is a general economic principle which suggests that new technologies usually only cause unemployment during a transitional period. Following this transitional period, higher levels of economic activity and employment are generated by the widespread application of the new technology. Of course there may be very real disruption and suffering during this "transitional phase".

The higher levels of economic activity and employment following the transitional phase are not usually a return to the employment patterns which existed before the technology was introduced. There will be new jobs and markets which are likely to be very different from those existing previously. This certainly seems to have happened in the case of Information Technology. Why would we expect AI to differ from this general pattern?

Some people certainly do think that AI is different. In the first industrial revolution machines replaced much manual work. In the information technology revolution machines replaced much routine administrative work. One does not, for example, see job advertisements for filing clerks in today's computerized societies. Many people see AI technology as about to replace more challenging intellectual work such as decision making, medical diagnosis, and maybe even teaching.

The first thing to say about this is that the present technology is not really any sort of employment threat to managers, doctors, and teachers. AI certainly makes a big contribution to modern management, medicine, and teaching but it has not yet caused any vast numbers of redundancies nor is it likely to for the foreseeable future.

A second and perhaps more interesting observation is that present-day and foreseeable AI technology seems more able to replace highly-specialized jobs than more general or human ones. Let's take the field of medicine as an example. We saw in chapter 2 how it was possible to build AI systems that could outperform human medical specialists, at least within their specialized area. If a medical professional relies on their detailed knowledge of a relatively narrow area for their employment and promotion – as do many consultant physicians – then they should be more worried about the threat from AI than a general practitioner, for example. A medical professional who uses a great deal of general knowledge and human interaction skills in her work is much less threatened by AI technology.

Lastly, it's obvious but worth saying at this point – we are not short of human intelligence. My personal belief is that all technology should improve life for humans, so replacing human intelligence is an unattractive development for me. It is also likely to be highly unprofitable and generally pointless as we have seen. There are many suitable applications for Artificial Intelligence which will enrich life for us humans and it is into those applications that it should be directed.

The truly golden scenario for the use of AI technology is that it will enable us humans to become effectively much more intelligent. Just as the mechanical digger vastly increased the amount of digging that one man could do, so the use of mechanical knowledge-manipulation tools – like data mining for example – can vastly increase the amount of intellectual digging that one man (or woman) can do. AI can act as an "intelligence amplifier" for us all. Using machines to help find and manipulate knowledge and ideas will make us all much cleverer.

There are some clouds in the way of this golden prospect, however. Human history shows a consistent tendency for those in power to want to prevent too much thinking by those over whom they exercise that power. Even the brief history of the IT industry suggests that it has mainly been used to empower executive and management activities at the expense of those lower down the structure. Data mining is a very powerful technology but it is more often used by those in power to draw up consumer and voter profiles than it is used by those consumers and voters. Many readers of this book might well be alarmed if they knew just how accurately their spending (and other) habits had been profiled. AI technology enables remarkably accurate predictions to be made from such profiles.

It would be good to have much more widespread public discussion of the use of AI technology. Because it is a powerful technology it often has a tendency vastly to increase the differences between those who have power and wealth and those who do not. This is true on a local scale, for example within a corporation. Here managers can monitor keyboard activity, phone conversations, emails, and so on of their employees. They can (and do) use data mining to profile customers, employees, and potential recruits. It is not clear that AI technology is benefiting customers or employees to the same extent. This may well be distorting the balance of power within corporations.

It is also true on a global scale where AI heightens the differences between those countries which possess this technology and those which do not. AI has always and continues to receive a major proportion of its financial support directly or indirectly from the military in a handful of affluent countries. The military are prepared to support AI research precisely because it is a powerful technology. Its contribution to modern military operations is largely unseen but very real. As with the applications described in chapter 2, AI can contribute to planning, logistics, communication, and decision support. The contribution of AI in these areas is very real and makes countries which employ it much more effective militarily than those who do not.

Like all technologies, AI can be employed in ways that have good social consequences and ways that have bad social consequences. Compared with its contemporaries such as nuclear fission and genetic modification, on the whole it seems rather unthreatening. Nevertheless there are important social issues to be considered – particularly the question of who gains from the technology and who loses. This should be the key question in public debate. It is urgent and real – unlike the more often discussed possibility of the machines taking over the world.

## will robots rule the world?

There have been many scare stories about robots taking over the world. Indeed the very word "robot" owes its existence to the first of these stories – by Carel Kapek in 1920. Where these stories are advanced as art then much of their function is to tell us about ourselves and our fears. When they are advanced as serious predictions then they are far more misleading.

I hope it is by now clear that none of the technologies discussed in this book are likely to be able to rule a filing cabinet, let alone the world for the foreseeable future. Even if they ever began to show that sort of ability we could just pull out the plug anyway. Why, then, is the myth of the robot takeover so persistent?

Firstly, it has to be said that some people see a robot takeover as very positive. Hans Moravec, for example, a prominent researcher who has been building robots at Carnegie Mellon

University since 1980, sees the supplanting of humans by robots later this century as largely positive. They are our "Mind Children", he says, and like good children they will help us prepare for a happy retirement. They will also become the brave initiative-takers and carry, at least the memory of, our human culture out into space (Moravec, 1990).

Moravec expects the future rate of progress in AI to be somewhat quicker than it was in the twentieth century. In the immediate future there does not seem to be much prospect of robots overtaking the intellectual capacities of humanity in the way in which he describes. In addition Moravec and other writers who talk of machines overtaking humanity often neglect the extent to which humans are a "moving target". These writers suggest that human intelligence has not evolved since the Stone Age and that is not strictly true. That is to say that humanity has adapted to agriculture, to industrial then postindustrial society, and so far shows every sign of being able to adapt to an age of intelligent machinery. Our technology is not something with which we compete. It is something with which we are in a mutually dependent and co-operative relationship. The biological word would be "symbiosis". One area where this can be seen happening is in the fascinating relationship between AI and art, which is briefly explored in the next section.

In the immediate future, therefore, though machines will undoubtedly get smarter, it does not seem that they will pose any great threat to humanity. People will tend to use AI to amplify their own human intelligence – as we often use IT nowadays. This may give certain groups and nations more power, but it will also enable most people to achieve far more than they can at present.

"What though of the distant future?" you might ask. Well if we make it distant enough then almost anything is possible but, even then, a robot takeover is still highly unlikely. People who advance such scare stories usually gloss over the process by which it happens. For example, it turns out that we can't actually pull out the plug because the robots get too smart too quickly. In some stories a desperate military organization builds overly destructive robots. Sometimes writers simply extrapolate the development of smarter and smarter machines and assume (rather stupidly) that humans will remain at their present level.

We can't really ignore the details of how a robot takeover might happen. The process by which this possible future robot takeover might happen is important because that will determine what we can do to prevent it. Once the proponents of such stories are pinned down on the details of how it could happen – a maverick researcher, a military necessity, human decadence – then it becomes obvious what we can and should do to prevent it. None of these things is inevitable. Humans can and should take steps to control risky developments in science. Maybe we need to examine and change the political control of military research. These are, however, political decisions about how we want to live and AI seems an innocent party in such decisions.

However, there's a much more important and convincing reason why nobody should be worried about robots ever ruling the world. It is simply not what they are designed to do. Now, many things really are designed to take over the world. Take daisies, for example. Their evolutionary history has programmed them to try to colonize all available space and they continue to do this. In a very real sense evolution has programmed daisies always to compete for more resources. Single-celled organisms such as bacteria are even more dangerous in that, not only are they programmed to take over all available space, they may well kill all humans in the process of doing so. In comparison, robots are not programmed or evolved to be any sort of threat. What is more, since we can understand and control robots a lot better then we can daisies and bacteria, we really have nothing to worry about.

Just because robots aren't about to take over the world doesn't mean that there is nothing serious to discuss about our relationship with them. In fact there is a large number of social, legal, and moral issues which are raised when we ask how people will live with clever (or even intelligent) machines.

One issue that is often discussed over coffee by people in AI and by some science fiction writers is perhaps the reverse of the robot takeover. This is the question of whether humans will systematically mistreat robots. If (and this is a condition that has not yet been met) we build machines that are capable of *genuine* suffering then it would be wrong to mistreat them. Again this doesn't seem to be an immediate problem and there are steps we could take to prevent it becoming a problem. The most important of these steps would probably be to make it illegal to build machines that had the capacity for genuine suffering. Some people think that it will not be as simple as this and that any machine that is genuinely intelligent will just have to be capable of real pleasure and suffering. Scientifically, we just don't know yet.

One thing that science has told us, though, is that humans can have a tendency to be very cruel, so perhaps this problem is more of a serious possibility than the robot takeover problem. Even if it turns out to be possible to build genuine suffering into some sort of machine and we then manage to stop people building such machines, there is still something worrying about passing a law that essentially makes it OK for people to mistreat robots. The worry is that we are effectively condoning rather cruel behaviour by humans. Similar worries apply to the use of AI in the sex industry, which is considered in the final section of this chapter.

Already millions of teenagers (and adults) shoot and kill computer-generated characters as a leisure activity. The computer games industry is more profitable than the Hollywood film industry and the continuing popularity of so-called "shoot-em-up" games shows that it has tapped into an important human need. The tendency towards increasingly violent and increasingly realistic computer games is in itself inherently worrying and AI will, almost certainly, be drawn into it. What is happening now in the computer games industry is that clever technologists are working to make more life-like opponents – often these days including AI – so that humans can kill and mutilate them. If this is normal human treatment, then the prospects for robots are not all that good.

## ai and art

There is, in Western culture, a false dichotomy between Art and Science. People often tend to see these two ways of looking at the world as very different when, in fact, they are rather similar. The interaction between AI and art provides an interesting illustration of this. There are a number of ways in which ideas (and technology) cross over between AI and art. Firstly artists use AI programs and robots to generate all sorts of art work. Secondly many AI researchers look at art in order to get a better understanding of how intelligence in general operates.

Both these crossovers have many dimensions. In the case of the first there have been many attempts to generate stories and poetry using AI programs. These programs tend to be knowledge-based. It may seem strange at first glance to think of creativity as based upon knowledge but this is one of the important ideas that flows in both directions across the supposed divide between art and science in this area. For example, Paul Hodgson, a professional jazz musician, wrote a program called Improviser. This program performs jazz improvisations in real time after the style of Charlie Parker. One thing that is interesting about this program is that it contains a knowledge-base of the most commonly used chord structures in Western music. "Isn't Jazz improvisation all about breaking the rules?" you might ask. Well yes it is, but it turns out that a knowledge of the rules is an essential ingredient of creatively breaking them.

Many people have written programs to generate poetry. I hope that, by this stage in the book, readers will see that the motivation for such work is not to replace human artists. More often it is way of exploring the complex rules which lie behind the way in which humans do this. I must admit that I am not usually moved by computer-generated poetry in the way that I am by some human-generated verse. However, I must also admit that the computer-generated verse is slowly getting better. I would make similar remarks about computer-generated stories.

There is an interesting question about to what extent, if any, we should credit the computer with *creating* these stories, poems, and jazz tunes. Perhaps all the credit should go to the programmers who have put in the rules. There is no simple answer to this question. On the one hand we could say that the programmers simply used the AI as a tool to enhance their own creativity. On the other, we could say that, for example, Hodgson simply taught Improviser to play. What it actually plays on any given occasion he cannot directly control. The truth probably lies in the difficult area in between. I believe that it is significant that the best AI art work almost always involves professional artists. However, it is also the case that these artists would usually say that they are not using the computer simply as an instrument. We can see here the beginnings of a fascinating symbiosis between humans and AI.

So far we have considered mainly the knowledge-based approaches to AI, but the rise of evolutionary computing has also seen the development of much art work based on evolutionary techniques. If programs such as genetic algorithms are used to generate patterns – say on canvas or in sound – they often have an exquisite beauty. One reason for this is that they are, to some extent, following the mathematics of natural life. Artists have started to explore many of these possibilities. In this case it may be even harder to say that the computer is merely an instrument, since the patterns themselves are evolved in a very similar way to natural evolution. An interesting twist to this is when artists evolve virtual creatures in a virtual environment and then use the activities of these creatures to create artistic patterns. In this case the human creator is even more remote from the finished work.

Many modern artists have also noticed AI and incorporated it into their work in exciting and thought-provoking ways. As well as being a recurring theme in science fiction, contemporary real AI is often included in novels. Novelists have used my workplace as a setting and my colleagues and I often read AI-based novels as much trying to spot personalities as gripped by the plot. However one of the most thought-provoking contemporary artists drawing inspiration from, and perhaps holding up a mirror to AI is an Australian performance artist called Stelarc.

Stelarc takes this notion of symbiosis with AI a stage further in that he has performed a number of installations that deliberately challenge our ideas of where flesh ends and technology begins. He has incorporated his own body into these installations in various ways.

My personal favourite is an installation in which he controlled a robot arm with electrical signals from his abdominal and leg muscles, whilst his other flesh arm was completely under the control of computer generated signals via a touch-screen stimulation system with electrodes attached to the skin (see photograph on p.120). Thus we see a mechanical arm under human control and a human arm under machine control.

## the future

Speculation on the future of science and technology is a famously dangerous thing to do. In the case of a relatively young area of science and technology such as AI it is particularly foolish. What I propose to do instead is to outline some of the work that is going on at present. Even this is likely to be incomplete because many people will be trying out new ideas that have not yet been publicized. An impenetrable wall of secrecy may surround some military AI research. By now readers

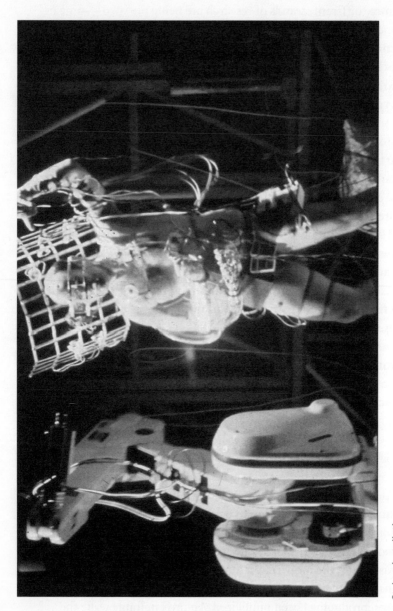

*Stelarc installation*

should have a clear enough general picture of AI to realize that so many different strands of research are being pursued – some in such small-scale ways – that even the best-informed commentators may miss things. There is a tremendous diversity of interest and effort within AI. Indeed my genuine hope is that some readers will be inspired by what they have read into exploring novel ideas for themselves.

## what's going on now?

Well, firstly, all the sorts of research described in chapters 1–4 are continuing. They may not all be the very height of fashion in AI – and AI is horribly fashion-conscious. Nevertheless let's be perfectly clear that they all work. Indeed many work so well that they are now generally considered to be more technology than science. Most people in advanced industrial societies use pieces of AI technology every day without even being aware of it. For example, all you see is that you call dial a number and get a connection on your mobile. The AI program that allocates routes according to demand is quite invisible. However, this is merely a sign that AI has matured as a technology. Good technology is usually invisible. In spite of what I said in chapter 2 most people in computing nowadays would see the technology of search, for example, as part of their own domain, rather than that of AI. My personal view is that all modern computer science has, as a matter of history, developed from AI. AI was there first.

When the first electronic computers were built, their designers often referred to them as "electronic brains". Indeed the very word "computer" meant a human who performed arithmetical tasks until the modern electronic computer became common. Turing and his associates at Bletchley Park and Princeton talked of reproducing human intelligent behaviour long before the modern IT industry took computing as a word for electronic filing and messaging.

AI has generated many ideas that have become part of computing in general. Time-sharing – the now almost universal technique for using a computer CPU (Central Processing Unit) to divide its working between several programs at once – originated in an AI lab. We noted in chapter 2 that fast prototyping of computer programs was developed in the knowledge-based systems area of AI. If you use a search engine on the web, you will probably be using AI technology or, at the very least, technology that has spun off from AI. Indeed some search engines are now so sophisticated that their working has been said to throw light on ways in which human memory might work.[1]

The scientific side to AI is also proceeding apace. The links with biology are being further strengthened in the field known as neuromorphic engineering. This is essentially the attempt to "reverse engineer" biological mechanisms. It is not easy and it is important that the engineering is done at the right level. At present this is taken to be at the level of a single neuron. This field holds out the promise of advancing both AI and biology tremendously. Similar observations apply to the links between AI and neuroscience.

Fundamental research is continuing (subject to fashion) in almost all areas of AI. One of the most fashionable areas at present is agent technology. An agent can be anything from a small program that, for example, represents your interests on the Internet or perhaps a fairly well drawn artificial character that helps you use a program or shoots back at you in a computer game.

Building situated, embodied robots is certainly continuing apace and some are finding useful application, such as maintaining sewers in German cities. Sophisticated robots are also needed for tasks such as exploring Mars. There will undoubtedly be new applications for such robots and this is an area where the applications can drive the theory. However, one important and immediate task is to find ways in which these various, very different local successes can be integrated with each other.

## integration

In chapter 4 reference was made to competing factions in AI. This has both positive and negative consequences. The most positive consequence is that people in AI have tackled the widest possible variety of subjects and techniques. Since the whole of human and animal behaviour is included in the scope of AI, together with robots and computers this is, perhaps, not surprising. It is very hard to convey in print just how exciting it is to be involved even in a small way in the beginnings of such a wide-ranging quest, but almost everybody in the field shares this sense of excitement to some degree.

The negative side was also remarked on in chapter 4. There is probably far too little in the way of attempts to integrate different approaches. Many people in AI see other approaches as having literally nothing to tell them and may even refuse to study them. This is almost certainly a mistake. As we have seen, intelligence is an immensely complex and poorly understood area. Many different techniques need to be employed simultaneously to research this area. That there should be any sort of theoretical separation between, say, the neural net approach described in chapter 3 and the knowledge elicitation work described in chapter 2 is obviously rather daft. What is worse, both these approaches are often roundly condemned by those who build situated robots.

Researchers may well not always benefit from looking at AI from a totally different perspective from their own. However, if they call these other perspectives "wrong" or "foolish" rather than simply "different" then they harm their own research. One reason for this is that it creates a totally unnecessary obstacle to borrowing from other ideas, from other methods, or sometimes even considering certain questions. Research into intelligent behaviour is difficult enough anyway. There really is no need to create extra obstacles like these.

Almost all of the different approaches to AI that have emerged over the years have yielded useful parts of the answer to some very difficult scientific questions. It is most certainly not a problem that *individually* they do not provide some sort of single complete solution. It has long been my view that what is needed in AI is a rejection of simplistic "gold standards" of intelligent behaviour. Intelligence is, as has been repeatedly observed throughout this book, complex and multi-dimensional.

Outside the academic world different approaches to AI cohabit not only in the same office but often in the same program. Clementine (described in chapter 2) would be a good example. Obviously integration is possible and, in this case at least, profitable too.

When I ask researchers why just so much effort goes into criticizing the efforts and views of others, one particular response occurs most frequently: It's necessary to get funding, they say. This may be the case in academia but it is not usually the practice in industry, where the question, "Does it work?" seems much more important than "Is it politically acceptable within our research paradigm?"

It is time for people to admit that any worthwhile account of intelligent behaviour has to include (at the very least) knowledge, neurons, and situatedness. Some philosophers are attempting to show at a rather abstract level how this can be done. We must hope that it will not be too long before this sort of integration of different approaches begins to pay off at a more practical level. And this practical pay off from integration may well characterize the immediate future progress of AI. It is no coincidence that the promising trends described at the end of this chapter involve some degree of integration of different approaches to AI.

Of course, integrating different approaches to AI requires

much more than simply adopting the right eclectic attitude. Because the different approaches look at the problem in fundamentally different ways, it is no simple task usefully to put them together. Let's look again at the problem with Cog. Rod Brooks laments that Cog cannot distinguish a glasses case from a cell phone. To him this is a serious failing of the research programme. To an AI researcher from the knowledge-based approach to AI, this problem is neither surprising nor difficult to fix. In knowledge-based terms Cog cannot distinguish the two items because it has no knowledge of either. An easy solution would be to program in some rules which capture the differences between phones and glasses cases.

A researcher using the connectionist approach would not program in rules but instead attempt to train an artificial neural net by showing examples. Stereotypical examples of cell phones and glasses cases would be presented in front of Cog's optical apparatus and the weights in the net modified until it reliably distinguished the two classes of object. There's no explicit rule in this case but the net is acting as a discrimination filter.

To Brooks and his team at MIT both of these methods would be an unthinkable piece of cheating. If they had wanted to design a device for telling phones from glasses cases there are many ways they could have done it. Phones usually weigh more than glasses cases so an arm on a spring calibrated to fall one way for a phone and the other way for glasses would probably do the job. To the MIT team this is just another way in which designers can cheat by using their knowledge of the world to build a specific solution to a specific problem into a machine. Programming in specific rules or training a net to do a very similar job to the rules are other ways. The whole point about Cog was that it was supposed to work this sort of thing out for itself.

But to the connectionist or the knowledge-based researcher, the MIT approach verges on wishful thinking. If and when Cog can distinguish cell phones from glasses cases, it must be on the basis of such a discriminatory rule. The rule may not be explicit, or even detectable, in its internal working but it is manifest in its behaviour. Simply building in good optical devices (and Cog has very good optical devices indeed), connecting these up to a powerful computer together with manipulators (arms) and so on and hoping for some interesting behaviours to emerge is not a valid research method. Even worse, when interesting behaviour does not emerge, as in this example, the temptation to tweak the robot a little to help things along becomes very great indeed.

So, to the connectionists and knowledge-based AI researchers, Cog's failure to distinguish cell phones from glasses cases is praiseworthy in that it shows that the team refrained from a simple cheat. It would be very easy to sneak in some small addition to part of a computer program to help Cog perform this task. On the other hand, they would probably also say that it just shows the total folly of this approach.

Perhaps this example helps to explain why AI researchers can sometimes be cynical about each other's methods. Integration of the various successful AI techniques will take more than pious words. There are many practical and philosophical problems to be overcome before integration of the different approaches can bear fruit. However, there are ways in which integration of the diverse sub-fields of AI is becoming central to research and the next section considers one of these – the field of agent technology.

## artificial agents

As was said in the introduction to this chapter, technically an agent can be many things. Firstly the area is too new for one stable definition to have emerged. Secondly, the deliberate crossing of boundaries has helped this field of AI combine ideas from other fields in a useful way. This approach to AI is often deliberately trying to combine some of the ideas of situated robots with other AI techniques. What is important about agents is not what they are but what they do.

Because agents are complete and relatively autonomous, they have to solve most of the problems of intelligent behaviour – things such as perceiving their environment and acting effectively in it – all at once. So, looking at AI from the point of view of how to build an agent to perform a particular task attempts to conform to at least some of the principles of situatedness and holism which were discussed in chapter 4.

One promising strand of research is the combination of large numbers of such agents. Although the intelligence of each individual may be limited, when they co-operate as a sort of "society" much more useful behaviour can emerge. Work on the behaviour of social insects such as ants has thrown light on some of the details of how comparatively complex tasks are achieved by combinations of relatively unsophisticated individual animals. It is hoped that communities of co-operating (and perhaps of competing) agents could also perform much more complex tasks than an individual agent could perform.

A good feature of agent-based technology is that it can be, in some examples, so simple that almost anybody can build it. The characters in computer games can be viewed as agents – particularly if they are capable of some autonomous decision making. Some types of program exploring the Internet – even the ones that send you unwanted mail – are examples of agent technology. The fact that this technology is relatively easy to build and relatively undisciplined means that progress is more likely to be made. Somebody somewhere will perhaps try the next big idea.

However, there are also examples of agent technology at the other end of the scale from small simple programs. Autonomous or semi-autonomous robots may often be examples of the same sort of technology. There are many useful application areas for such robots: take exploring Mars, for example. A robot moving about the surface of Mars can't simply be steered by a human controller like the robots in BattleBots or Robot Wars. The delay in receiving a radio signal from Mars and then in sending the control signal back could be several minutes (between 9 and 48 minutes to be precise) – during which time the robot may well have run into trouble. The robot has to use "on-board intelligence" to avoid falling into large holes.

Please remember that the semi-autonomous robot exploring the surface of Mars is just the tip of the AI iceberg. There will be three probes sent to Mars during 2003 alone. These probes depend on a collection of many different sorts of AI technology. There are mission-scheduling systems which plan the technical requirements. There are knowledge-based systems which the scientists use to predict interesting places to explore on Mars. Knowledge-based systems have been very successful in predicting the locations of mineral deposits on Earth. AI software will also be used to enhance the images sent back and in decision-support systems to help controllers during the mission.

Another interesting application is in the maintenance of sewer systems. Here mechanically relatively simple robots that propel themselves through underground pipes can help clear blockages. However, should an individual robot get stuck it can ask for assistance from other robots nearby, before asking for human intervention. This sort of co-operative behaviour from semi-autonomous robots overlaps with other aspects of agent technology. For example, the opponents that fire back at the player in "shoot-em-up" computer games may co-operate in ways that are, in principle, the same as the sewer-maintenance robots. Even though, in the case of the computer game, they are actually nothing more than pieces of computer software that are generating images on the screen.

## virtual girlfriends and artificial companions

Thinking of agents as computer software that generates images on the screen leads to some more very interesting applications for AI. Already major Hollywood film companies are proposing to make films with computer-generated stars. Of course, artificial characters in the movies date back at least to Mickey Mouse's famous 1928 appearance. The latest technology offers something even more spectacular – computer-generated characters that are indistinguishable on-screen from human actors. Plans are already well advanced to allow Marilyn Monroe and Humphrey Bogart (or at least computer-based reconstructions of them) to begin making movies again. I have little doubt that the technical problems will soon be overcome. This sort of technology is expensive but not scientifically difficult in the way that much AI research is. Since human actors are extremely expensive (at least in Hollywood), the financial incentive to use computer-generated stars instead is very real.

Combining Hollywood's ability to generate totally convincing human-like images and the AI technology of autonomous or semi autonomous agents raises further interesting possibilities for future developments. One industry that could be very interested in these developments is the sex industry. I have already mentioned that this application area is one (and probably the only) place where deliberate imitation of human characteristics is a research goal. Looking at the area relatively dispassionately, there are a number of possible applications for AI here.

It is important to remember that human sexual drives have often shaped technological developments. For example, the videotape standard which came to dominate the domestic market (VHS) did so mainly because it was the standard adopted by the pornographic industry. The history of the world-wide web is similar. The Internet was originally a military development used by a few academics, initially because they were connected with military research. The world-wide web is an interface to the Internet which was developed by nuclear scientists to allow them easily to view each others' research. The technology only took off and entered everyday use because it was the ideal technology for passing pornographic images around the world. Even today, this is a major (if not the major) use of the world-wide web.

One conclusion that might be drawn from these previous technologies is that there is a tremendous demand for this sort of application of technology. The possibility of a large market and large profits are reasons why these possibilities need to be taken seriously. AI will be drawn into this industry in various ways. One may be the extension of the computer-generated artificial characters that exist in computer games. As has already been observed, these characters nowadays usually contain AI. Whether they exist primarily to be shot, as in the most popular computer games, or they exist to be nurtured as in the other main genre of such games, they are, at present, very much cartoon characters. Combining this AI technology with the ability to generate completely realistic artificial characters will be an opportunity that the sex industry is unlikely to miss.

Already there is a strong demand for virtual girlfriends – that is, computer-based characters that perform at least some of the roles of a female companion. (It seems to be most trendy among Japanese businessmen though I forebear to speculate on why this might be.) Combination of existing technologies might well allow the creation of relatively realistic artificial companions in the relatively near future. Screen-based characters are more likely than physical embodied robots for the present. Of course, that is precisely what most current film and TV stars are.

It is very easy to experience an instant emotional reaction to this possibility. However, it is only a combination of existing technological developments and existing social trends. Some writers, such as Neil Frude, see it as a truly dystopian prospect. People will prefer the artificial to the real, they claim. They will much prefer to stay at home with their artificial companions than to venture out into the many risks of the real world. This will have the result that normal human interaction with all its unpredictability and difficulties will decline.[2] Against this point of view one might argue that the screen-based heart-throbs of the twentieth century did not dissolve human society. Most people can distinguish the virtual from the real and people will adapt to this technology just as they have adapted to its ancestors such as TV and computer games. However, the twentieth-century virtual technologies – film, TV soaps, and computer games – often generated vast profits, so it is a financially attractive application area.

Not all artificial companions are likely to be developed by the sex industry. An interesting military application of this technology is known by its developers as "The Pilot's Associate". Because pilots of modern combat aircraft can be extremely overloaded, a knowledge-based system was developed which could constantly direct their attention to the most urgent matter. This might be an incoming missile, an overheating engine, low fuel, or some other pressing problem. Research showed that the most effective way to present this urgent information to the human pilot was in the form of an insistent female voice in their headphones. This probably explains why the pilots' name for the system is "Bitching Betty".

A promising and socially beneficial application area for artificial companions is in systems (and this includes physical robots) that look after elderly people in various ways. These can include integrated domestic technology that monitors such things as cooking, bathwater temperature, and so on to help prevent accidents. Robots can provide physical assistance with various domestic chores. In Europe and the US where there are ageing and affluent populations such technology has a large and receptive market. Whether this technology will be developed in parallel with the sex industry's use of AI or one will spin off from the other is an interesting question and I expect to see it answered relatively soon.

However, if I am forced to speculate on the next big development in AI, I would have to say that it is likely to take most people in the field completely by surprise. I certainly expect to be surprised. Previous developments have often appeared to come out of nowhere. Small groups of researchers who have been looking at familiar problems in a different way sometimes produce a breakthrough. Sometimes researchers go back to a long-neglected idea and try it again using more modern computers. Sometimes a maverick just comes up with a revolutionary idea and it turns out to work.

As I said in the preface, it's an exciting field to be in and it's an exciting time to be in it.

## further reading

One place to pursue the social implications of AI further is in *Reflections on AI: the Legal, Moral, and Ethical Dimensions* (Whitby, 1996).

Hans Moravec makes the case for robots taking over the world in *Mind Children* (1990).

Margaret Boden has written a book (*The Creative Mind*, 1990) which applies AI to artistic (and other) creativity. This has become the starting point for just about anybody who is interested in the relationship between AI and Art.

Stelarc has a website at: http://www.stelarc.va.com.au/ This gives a taste of his work and thinking.

## notes

1. Clark, 'Local Associations and Global Reason', pp. 115–40.
2. Frude, *The Intimate Machine*.

# bibliography

Abelson, R.P., 'The Structure of Belief Systems', in R.C. Schank and
K.M. Colby, (eds) *Computer Models of Thought and Language*.
San Francisco, W.H. Freeman, 1973.

Aleksander, I. and Morton, H., *An Introduction to Neural
Computing*. London, Chapman & Hall, 1990.

Anderson, A.R. (ed.), *Minds and Machines*. Englewood Cliffs, NJ,
Prentice-Hall, 1964.

Asimov, I., *I Robot*. St. Albans, Panther Books Ltd, 1968.

Beerel, A.C., *Expert Systems Strategic and Implications Applications*.
Chichester, Ellis Horwood, 1987.

Boden, M.A., *Minds and Mechanisms, Philosophical Psychology and
Computational Models*. Brighton, Harvester, 1981.

—, 'Impacts of Artificial Intelligence', *AISB Quarterly*, 49, Winter
83–84.

—, *Artificial Intelligence and Natural Man* (2nd edn). London, MIT
Press, 1987.

—, *The Creative Mind, Myths and Mechanisms*. London, Weidenfeld
& Nicolson, 1990.

Boden, M.A. (ed.), *Dimensions of Creativity*. Cambridge, MA and
Bradford, MIT, 1994.

Born, R.P. (ed.), *AI, The Case Against*. London, Croom Helm, 1987.

Bramer, M.A. (ed.), *Research and Development in Expert Systems III*.
Cambridge, Cambridge University Press, 1987.

Brooks, R.A., 'Intelligence Without Representation', *Artificial
Intelligence Journal*, 47, 1991, pp. 139–59.

—, *Robot: The Future of Flesh and Machines*. London, Penguin, 2002.

Ciampi, C. (ed.), *Artificial Intelligence and Legal Information Systems*. Oxford, North-Holland, 1982.

Clark, A., *Microcognition: Philosophy, Cognitive Science, and Parallel Distributed Processing*. London, MIT Press, 1989.

—, *Being There, Putting Brain, Body, and World Together Again*. London, MIT Press, 1997.

—, 'Local Associations and Global Reason: Fodor's Problem and Second-Order Search', *Cognitive Science Quarterly*, 2002, 2, 115–40.

Colby, K.M., Hilf, F.D., Weber, S., and Kraemer, H.C., 'Turing-Like Indistinguishability Tests for the Validation of a Computer Simulation of Paranoid Processes', *AI*, 3, 1972.

Crane, T., *The Mechanical Mind, A Philosophical Introduction to Minds, Machines, and Mental Representation*. London, Penguin, 1995.

Dennett, D., *Brainstorms*. Brighton, Harvester, 1978.

—, 'Cognitive Wheels' (1984), in *Brainchildren, Essays on Designing Minds*. London, Penguin, 1998.

—, *Consciousness Explained*. London, Penguin, 1993.

—, *Darwin's Dangerous Idea, Evolution and the Meanings of Life*. London, Penguin, 1995.

—, *Brainchildren, Essays on Designing Minds*. London, Penguin, 1998.

Derry, T.K. and Williams, T.I., *A Short History of Technology*. Oxford, Oxford University Press, 1960.

Dreyfus, H.L., *What Computers can't do: a Critique of Artificial Intelligence*. New York, Harper & Row, 1972.

Dreyfus, H.L. and Dreyfus, S., *Mind over Machine*. New York, Free Press, 1986.

Enver, T., *Britain's Best Kept Secret: Ultra's Base at Bletchley Park*. Stroud, Glos., Alan Sutton, 1994.

Feigenbaum, E.A. and McCorduck, P., *The Fifth Generation*. London, Pan Books, 1984.

Forester, T. (ed.), *The Information Technology Revolution*. Oxford, Blackwell, 1985.

Forsyth, R. and Naylor, C., *The Hitchhiker's Guide to Artificial Intelligence*. London, Chapman & Hall, 1986.

Frude, N., *The Intimate Machine*. New York, New American Library, 1983.

Gardner, A., *An Artificial Intelligence Approach to Legal Reasoning*. London, MIT Press, 1987.

Gill, K.S. (ed.), *Artificial Intelligence for Society*. Chichester, John Wiley, 1986.

Grand, S., *Creation: Life and How to Make it*. London, Weidenfeld & Nicolson, 2000.

—, *Growing up with Lucy* (forthcoming). London, Weidenfeld & Nicolson, 2003.

Harnish, R.M., *Minds, Brains, and Computers, An Historical Introduction to the Foundations of Cognitive Science*. Oxford, Blackwell, 2002.

Harre, R., *Cognitive Science, A Philosophical Introduction*. London, Sage, 2002.

Hodges, A., *Alan Turing, the Enigma of Intelligence*. London, Unwin Paperbacks, 1985.

Hofstadter, D.R., *Godel Escher Bach, An Eternal Golden Braid*. Brighton, Harvester, 1977.

Hofstadter, D.R., and Dennett, D.C. (eds), *The Mind's I*. Brighton, Harvester, 1981.

Jackson, P., *Introduction to Expert Systems* (2nd edn). Wokingham, Addison-Wesley, 1990.

Kuhn, T.S., *The Structure of Scientific Revolutions*. University of Chicago Press, 1970.

LaChat, M., 'Artificial Intelligence and Ethics: An Exercise in the Moral Imagination', *The AI Magazine*, Summer 1986, pp. 70–9.

Leibnitz, G.W. von, *Monadology*. First published in English, Oxford, Clarendon Press, 1898.

Levy, S., *Artificial Life, The Quest for a New Creation*. London, Penguin, 1993.

Lighthill, J., et al., *Artificial Intelligence: a Paper Symposium*. London, Science Research Council, 1973.

Luger, G. and Stubblefield, W., *Artificial Intelligence Structures and Strategies for Complex Problem Solving* (2nd edn). Redwood City, CA, Benjamin/Cummins, 1993.

McCorduck, P., *Machines Who Think*. San Francisco, W.H. Freeman, 1979.

Michie, D., 'The Social Aspects of Artificial Intelligence', in T. Jones, (ed.), *Microelectronics and Society*. Milton Keynes, Open University Press, 1980.

Miller, G., *The Mating Mind: How Sexual Choice Shaped the Evolution of Human Nature*. London, William Heinemann, 2000.

Miller, P.L., *A Critiquing Approach to Expert Computer Advice: ATTENDING*. London, Pitman Publishing, 1984.

Minsky, M, *The Society of Mind*. New York, Simon & Schuster, 1986.

Moravec, H., *Mind Children, The Future of Robot and Human Intelligence*. Cambridge, MA, Harvard University Press, 1990.

Nagel, T., 'What is it like to be a bat?', *Philosophical Review* LXXXIII 1974 (repr. in *Mortal Questions*. Cambridge, Cambridge University Press, 1979).

—, *Mortal Questions*. Cambridge, Cambridge University Press, 1979.

O Nuallain, S., *The Search for Mind, A New Foundation for Cognitive Science*. Bristol, Intellect, 2002.

Owen, K., *Report on the Social Implications of Expert Systems and Artificial Intelligence*. BCS Publications, 1985.

Penrose, R., *The Emperor's New Mind: Concerning Computers, Minds, and the Laws of Physics*. Oxford, Oxford University Press, 1989.

Pfeifer, R. and Scheier, C., *Understanding Intelligence*. London, MIT Press, 1999.

Randell, B. (ed.), *The Origins of Digital Computers*. New York, Springer-Verlag, 1982.

Rich, E., *Artificial Intelligence*. New York, McGraw-Hill, 1983.

Russell, S. and Norvig, P., *Artificial Intelligence: A Modern Approach* (2nd edn). Upper Saddle River, NJ, Prentice Hall, 2003.

Searle, J., 'Minds, Brains and Programs', in *The Behavioral and Brain Sciences*, vol. 3, 1980; also in D.R. Hofstadter and D.C. Dennett, (eds), *The Mind's I*, Brighton, Harvester, 1981.

—, *Minds, Brains, and Science*. London, Penguin, 1991.

—, *The Rediscovery of the Mind*. London, MIT Press, 1994.

Shortcliffe, E.H., *Computer-based Medical Consultations: MYCIN*. New York, American Elsevier, 1976.

Sloman, A., *The Computer Revolution in Philosophy*. Brighton, Harvester, 1978.

Susskind, R., *Expert Systems in Law*. Oxford, Oxford University Press, 1987.

Taube, M., *Computers and Common Sense*. New York, Columbia, 1968.

Tennekes, H., *The Simple Science of Flight: From Insects to Jumbo Jets*. Cambridge, MA, MIT Press, 1997.

Thorton, C. and du Boulay, B., *Artificial Intelligence through Search*. Bristol, Intellect, 1992.

Torrance, S. (ed.), *The Mind and the Machine: Philosophical Aspects of Artificial Intelligence*. Chichester, Ellis Horwood, 1984.

Torrance, S., 'Ethics, Mind and Artifice', in K.S. Gill, (ed.), *Artificial Intelligence for Society*. Chichester, John Wiley, 1986.

Turing, A.M., 'Computing Machinery and Intelligence', first published in *Mind* LIX, 236; also in A.R. Anderson, (ed.), *Minds and Machines*. Englewood Cliffs, NJ, Prentice-Hall, 1964; and D.R. Hofstadter and D.C. Dennett (eds), *The Mind's I*. Brighton, Harvester, 1981.

Turkle, S., *The Second Self, Computers and the Human Spirit*. London, Granada, 1984.

Weinberg, G.M., *The Psychology of Computer Programming*. New York, Van Nostrand Rheinhold, 1971.

Weizenbaum, J., 'ELIZA – a computer program for the study of natural language communication between man and machine', Communications of the *ACM* 9 (1), 1966 pp. 36–45.

—, *Computer Power and Human Reason*. Harmondsworth, Penguin, 1984.

Whitby, B., 'AI: Some Immediate Dangers', in M. Yazdani and A. Narayanan (eds), *Artificial Intelligence: Human Effects*. Chichester, Ellis, Horwood, 1984.

—, 'The Computer as a Cultural Artefact', in K.S. Gill, (ed.), *Artificial Intelligence for Society*. Chichester, John Wiley, 1986.

—, *Artificial Intelligence: A Handbook of Professionalism*. Chichester, Ellis Horwood, 1988.

—, *Reflections on Artificial Intelligence, The Legal, Moral, and Ethical Dimension*. Bristol, Intellect, 1996.

Wood, S., 'Expert Systems for Theoretically Ill-formulated Domains', in M.A. Bramer, (ed.), *Research and Development in Expert Systems III*. Cambridge, Cambridge University Press, 1987.

—, *Planning and Decision-Making in Dynamic Domains*. Chichester, Ellis Horwood/Wiley, 1993.

Yazdani, M. and Narayanan, A. (eds), *Artificial Intelligence: Human Effects*. Chichester, Ellis Horwood, 1984.

Yazdani, M. and Whitby, B., 'Artificial Intelligence: building birds out of beer cans', *Robotica*, 5, pp. 89–92.

# index